Many aspects of human activity involve energy transfer of some type. *Human energetics in biological anthropology* considers various ways in which measurements of energy intake, expenditure and balance have been used to study human populations by biological anthropologists and human biologists. Central to this approach is the concept of adaptation and adaptability, placed in an ecological context by considering such processes in traditional subsistence economies in the developing world.

This volume will be useful for students and their teachers in biological anthropology, human population biology, nutritional anthropology and nutrition in the developing world.

T0275874

Cambridge Studies in Biological Anthropology 16

Human energetics in biological anthropology

Cambridge Studies in Biological Anthropology

Series Editors

G. W. Lasker
Department of Anatomy, Wayne State University,
Detroit, Michigan, USA

C. G. N. Mascie-Taylor
Department of Biological Anthropology,
University of Cambridge

D.F. Roberts
Department of Human Genetics,
University of Newcastle-upon-Tyne

R.A. Foley
Department of Biological Anthropology,
University of Cambridge

Human energetics in biological anthropology

STANLEY J. ULIJASZEK

Department of Biological Anthropology,
University of Cambridge, Cambridge, UK

CAMBRIDGE UNIVERSITY PRESS
Cambridge, New York, Melbourne, Madrid, Cape Town, Singapore, São Paulo

Cambridge University Press
The Edinburgh Building, Cambridge CB2 2RU, UK

Published in the United States of America by Cambridge University Press, New York

www.cambridge.org
Information on this title: www.cambridge.org/9780521432955

First published 1995
This digitally printed first paperback version 2005

A catalogue record for this publication is available from the British Library

Library of Congress Cataloguing in Publication data

Ulijaszek, Stanley J.
 Human energetics in biological anthropology / Stanley J. Ulijaszek.
  p. cm. – (Cambridge studies in biological anthropology; 16)
 Includes bibliographical references (p.) and index.
 ISBN 0 521 43295 2 (hardback)
  1. Physical anthropology. 2. Basal metabolism. 3. Energy
 metabolism. I. Title. II. Series.
 GN223.U45 1995
 573–dc20   94–31148 CIP

ISBN-13 978-0-521-43295-5 hardback
ISBN-10 0-521-43295-2 hardback

ISBN-13 978-0-521-01852-4 paperback
ISBN-10 0-521-01852-8 paperback

Contents

Preface

Nature is never simple, human nature less so. Energetics approaches have been used in biological anthropology to examine aspects of the human biocultural condition. Their use has a fairly recent and varied history, proving more successful in studies of human adaptation and adaptability than in broader ecological designs. Although much energetics-related research has considered physiological adaptation, new questions in the adaptability paradigm must consider the influence of behaviour and cultural change on human survival. In particular, behaviours relating to the determination of household size, the scheduling of subsistence tasks and the nature and degree of cooperative behaviour in relation to food production all profoundly influence human population biology and are subject to change in response to environmental, technological and cultural changes.

In recent years, a wealth of knowledge about the energy costs of pregnancy, lactation and physical growth, the ecological and physiological correlates of body size, and nutritional factors influencing human fertility has accumulated from the extensive use of energetics methods outside the laboratory and in the places where people live. This book reviews the human energetics approach and outlines a number of related debates in the anthropological arena that have been informed by such knowledge.

A number of people have been instrumental in guiding and developing my thinking on the relationships between energetics, adaptation, nutrition and anthropology, notably the late Derek Miller, Nick Norgan, Peter Heywood, John Lourie and Simon Strickland. I thank them.

Stanley J. Ulijaszek
Cambridge

1 *Introduction*

The use of energetics in biological anthropology began with the ecosystemic approach but has been used in less holistic ways to examine processes of human adaptation and adaptability. Ecosystematics attracted anthropologists, largely because it allowed holistic studies of humans in their environment (Moran, 1990) and suggested the possibility of common principles in biology and anthropology (Winterhalder, 1984). Ambitious in scope, many studies employing such techniques failed to match the claims made for them (Burnham, 1982). In particular, although biological anthropology seeks to elucidate the causes of variation within- and between-human populations, systems ecology has rarely ventured into causal explanations. At best, the ecosystem concept offers a macro-scale descriptive frame for the study of human ecology (Smith, 1984). However, such descriptions have helped to improve the understanding of subsistence-related processes in human population biology. More limited use of energetics within the human adaptability framework has provided insights into some of the processes that lead to human population variation (e.g. James, 1988; Waterlow, 1990a; Ellison, 1991; Bailey *et al.*, 1993; Ulijaszek, 1993).

In this volume, energetics approaches and the issues that can be addressed with them are examined, acknowledging that the adaptability approach is nested in the ecosystemic one. The adaptability approach has ceased to be practised to any significant degree, but it is argued that data collected in the past can continue to inform the understanding of human adaptive processes, especially when used in conjunction with newer information. In this chapter, the studies of ecosystematics, human adaptation and adaptability are outlined and related to energetics. An historical background is given, with examples of the type of work carried out, where appropriate.

Ecological energetics

Although the German biologist Ernst Haeckel is credited with being the first to use the term ecology (Haeckel, 1868), the science of ecology did not get underway until the turn of the twentieth century. Ecology has been

1

defined as the study of the interrelationships between living organisms and their environment (Lincoln, Boxshall & Clark, 1982). Despite the obvious implications of ecology for understanding human biology and behaviour, anthropologists were attracted to this approach only in the 1920s, when the term 'human ecology' was first used by the geographer H. H. Barrows (1922). In this and subsequent formulations (Adams, 1935; Park, 1936; Hawley, 1950) there was little attention paid to the causes and consequences of energy use, while adaptation was expressed in terms of social competition rather than biological function and Darwinian fitness.

The term 'ecosystem' was formally proposed by Tansley as a general term for the total complex of interacting organisms and their habitat. Through their interactions, the entire system is maintained (Tansley, 1946). These interactions involve a relatively stable set of relationships in which material, information and energy are in continuous circulation. The American biologist Lindeman (1942) focussed on the fixation of energy in ecosystems and the quantitative relations that must exist between different users of energy as it is spread around various populations of organisms within an ecosystem. This work helped to consolidate ecology as a discipline and was a major influence on anthropologists interested in understanding the relationships between human groups and their environments (Ellen, 1982).

Subsequently, the American anthropologist Julian Steward developed a theoretical framework, which he called cultural ecology (Steward, 1955), the first appropriation of ecological principles in anthropology. Stewart was convinced that the natural environment, through its effects on human subsistence behaviour, had a strong effect on the types of social and political structure which developed in societies that used comparable natural habitats. He was concerned with the relationships between environment and productive technology, the patterns of behaviour involved in resource acquisition through specific technologies in particular areas and the extent to which behaviour patterns involved in such acquisition influence other aspects of human behaviour. However, the quantification of these relationships was not a major concern.

Human ecologists have used a variety of approaches in their attempts to understand human group function and reproductive success. These have included: (1) demography; (2) the importance of culture and ritual in subsistence-related decision making; (3) economic and exchange relationships; and (4) energetics.

Demographic measures include vital statistics such as births and deaths at various stages of life, as well as fertility levels and in- and out-migration. If the aim is to measure fitness, then demography must be considered in the context of other factors, including food-getting ability, social stratification and economics, and knowledge and power. Anthropologists have examined the human ecology of ritual (Rappaport, 1968), warfare (Rappaport, 1968; Ferguson, 1989), trade (Thomas, 1976), foraging strategies (Hill *et al.*, 1984; 1985), and technological change (Bayliss-Smith, 1977).

Although the exchange value of materials and the relationships that ensue from exchange and distribution have long been the concern of economic anthropology (Seymour-Smith, 1986), they can, at best, only give a partial understanding of human ecosystemic relationships. Similarly, the study of information control, exchange and use in relation to subsistence strategies can lead only to partial knowledge of ecosystemic regulation (Moran, 1982). One way in which the biological factors involved in ecosystemic regulation can be understood is in terms of energy exchanged, used, created and stored in various forms. Ecological energetics could not develop until practical problems of the measurement of energy, in its many forms, were resolved.

Adaptation and adaptability

Adaptation and adaptability are central concepts in biological anthropology because they are the processes whereby beneficial relationships between humans and their environment are established and maintained. They are also the processes that allow change or accommodation to new conditions or circumstances. Adaptation has been considered from four perspectives: genetic, physiological, behavioural and cultural (Ellen, 1982; Harrison, 1993).

Genetic adaptation takes place through selection of the genotype, the genetic structure of the population being shaped by differential fertility and mortality. Physiological adaptation involves the shorter-term changes which individuals show in response to any of a variety of environmental stressors, among them low food availability. Behavioural adaptation includes types of behaviour that can confer some advantage, ultimately reproductive. Such behaviours may include proximate determinants of reproductive success, including patterns of resource acquisition, especially food and energy. Cultural adaptation involves the transmission of a body of knowledge and ideas, objects and actions being the products of those ideas. Although cultural structures can evolve as adaptive systems in response to environmental factors, not all aspects of culture can be assigned adaptive significance (Morphy, 1993).

The four types of adaptation do not exist in isolation from each other. Rather, they are linked across time and feed back on each other (Fig. 1.1). Genetic adaptation takes place across generations, as do many aspects of cultural adaptation. In this sense, genetic and cross-generational cultural

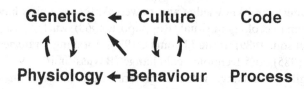

Figure. 1.1.   Codes and processes in human adaptation.

adaptation can be regarded as codes. Adaptation as process arises out of adaptation as code and includes physiological and behavioural adaptation. The physiological processes have a basis in genetics, while the processes that have a basis in culture are the behaviours that operate within the lifespan of the individual. They range from the extremely short-term behaviours, such as instantaneous decisions, to longer-term behaviours, such as the choice of marriage partner. If either physiological or behavioural processes serve to enhance reproductive success, either directly or indirectly, then they can become fixed in the genetic and cultural codes, respectively.

Thus there is feedback between processes and codes: physiological processes operating within the lifespan feed back on genetic code, influencing genetic adaptation, while behavioural processes, also taking place within the lifespan, feed back on cultural code. Behaviour and culture also influence genetic adaptation through, for example, kinship patterns and marriage laws which may affect differential reproductive success and biologial population structure through assortative mating. Furthermore, behaviours may influence physiological processes, while the diffusion of culture that has not evolved cross-generationally but has been adopted from another group or population is of increasing importance in a world in which the transmission of information across traditional barriers is ever increasing.

Although there is considerable overlap between definitions of human adaptation and adaptability (Ulijaszek, 1995), adaptability does not overlap with genetic adaptation. Furthermore, adaptability is also the ability to adapt. However, to be of some analytical value, it is important to define the limits of this ability, whether it be physiological, behavioural or cultural adaptation. From Fig. 1.1, it is clear that there are adaptive processes that take place across generations, and processes that occur within the lifespan but have influence on, or drive, cross-generational processes. The term adaptability has from the outset been reserved for the kind of responses that individuals make to changes in their environment that facilitate their survival and reproduction and is thought of as a

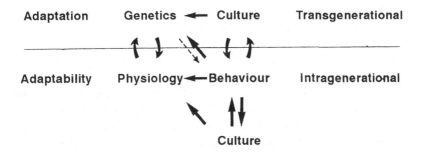

Figure 1.2. Relationships between human adaptation and adaptability.

property of an extant group (Harrison, 1993). Figure 1.1. can be redrawn to include human adaptability as a within-generational process (Fig. 1.2) and adaptation as a cross-generational process. Thus, human adaptability includes physiological and behavioural processes, as well as the adoption of cultural factors that may be of adaptive significance from other populations. There may also be behavioural adaptability in the choice of what cultural factors are adopted.

For example, physiological responses to low dietary energy availability include weight loss and body composition changes, as well as possible down-regulation of basal metabolism. Once body size matches energy resources, homeostasis is regained at a lower level of intake. However, a behavioural response might be to reduce physical activity. These are not mutually exclusive, and the mix of physiological and behavioural adaptability is constrained by cultural and genetic codes and states nested in both higher and lower levels of organisation than that of the individual.

At a lower level, the physiological state of different organs and tissues will determine the ability of the individual to undergo weight loss without functional impairment. In the maintenance of individual physiological homeostasis, there are circumstances in which a reduction in physical activity may be preferable to weight loss. This may, however, be in conflict with strategies suggested at higher levels of organisation, such as the group or community. For example, the need to perform arduous time-limited seasonal tasks such as planting or harvesting of crops to ensure food supplies for the coming year may rule out the possibility of reduced activity in the face of low food availability.

Energetics

Many aspects of human activity involve energy transfer of one sort or another (Harrison, 1982), and since the early 1960s there has been considerable attention paid by anthropologists to the energetics of human ecology Lee, 1965; Rappaport, 1968; Montgomery & Johnson, 1977; Little & Morren, 1977; Morren, 1977; Ellen, 1982; Ohtsuka, 1983). In particular, estimates of energy intakes, total expenditures, costs of activity and balances and flows have been used in attempts to understand human subsistence within the adaptation and adaptability framework (Bayliss-Smith & Feachem, 1977; Pimentel & Pimentel, 1979; Bayliss-Smith, 1982a,b; Thomas, Gage & Little, 1989; Ulijaszek & Strickland, 1993a). Concern is often with how the need for dietary energy and the ways in which it is obtained affect different aspects of human population biology or ecology (Haas & Pelletier, 1989; Thomas, McRae & Baker, 1982; Thomas *et al.*, 1989; Weitz *et al.*, 1989), or the implications of different subsistence and foraging strategies for fertility and biological fitness (Smith, 1979; Bertes, 1988; Hill & Kaplan, 1988a,b).

More recently, this has been extended to consider human responses and adaptations to seasonal energetic stresses (Dugdale & Payne, 1986; Brenton, 1988; de Garine & Harrison, 1988; Hitchcock, 1988; Huss-Ashmore & Goodman, 1988; Huss-Ashmore & Thomas, 1988; Little, Galvin & Leslie, 1988; Messer, 1988, 1989; Nyerges, 1988; Wheeler & Abdullah, 1988; Abdullah, 1989; Lawrence *et al.*, 1989; Payne, 1989; Ferro-Luzzi, 1990b; Ferro-Luzzi *et al.*, 1990; Thomas & Leatherman, 1990), and the role that energetics plays in reproductive ecology (Rosetta, 1990, 1994; Ellison, 1991; Bailey *et al.*, 1992).

Energy is an interconvertable currency that can be used in quantitative analysis of activities of human groups that are minimally or only partially involved in the cash economy. Thus, although the approach may not lead to a comprehensive understanding of human adaptation (Smith, 1979; Burnham, 1982), it can give an extensive account of that aspect of human functioning related to resource acquisition, subsistence and ecological and reproductive success (Ellen, 1982; Thomas *et al.*, 1982).

It is generally assumed that energy is the limiting factor in meeting the subsistence needs of most human populations, and if energy needs are met, the need for all other nutrients will also be met. This is not necessarily true, particularly in relation to iron and vitamin A intake. Furthermore, protein adequacy may be low when the staple food is low in protein. In practice, very few of the world's staple foods are low enough in protein to suggest likely deficiency in adults, even at low activity levels (Table 1.1). The safe level of protein intake, as defined by FAO/WHO/UNU (1985), is 0.75 g/kg of body weight. For adult males weighing 55 kg, the safe level of protein intake is 41 g. Of the 15 staple foods shown in Table 1.1, ten would supply this level of protein intake at all levels of physical activity (and therefore of

Table 1.1. *Protein content of staple foods and protein intakes of an hypothetical adult male, weighing 55 kg, at 1.4, 1.8 and 2.2 times basal metabolic rate (representing an inactive, moderately active, and very active person)*

Staple food	Energy (kJ/100 g)	Protein (g/100 g)	Protein intake (g) at		
			1.4 × BMR	1.8 × BMR	2.2 × BMR
Rye (*Secale* sp.)	1397	12.8	81.6	104.9	128.2
Oats (*Avena* sp.)	1565	13.1	74.6	95.9	117.2
Wheat (*Triticum* sp.)	1389	11.6	74.4	95.7	116.9
Millet (*Pennesetum* sp.)	1402	11.4	72.5	93.2	113.9
Barley (*Hordeum* sp.)	1368	10.5	68.3	87.8	107.3
Sorghum (*Sorghum* sp.)	1431	10.0	62.3	80.1	97.9
Maize (*Zea* sp.)	1460	9.1	55.5	71.4	87.2
Potato (*Solanum* sp.)	343	2.0	51.9	66.7	81.6
Taro (*Colocasia* sp.)	393	2.2	49.8	64.0	78.3
Rice (*Oryza* sp.)	1481	7.6	45.7	58.8	71.8
Yam (*Dioscorea* sp.)	427	1.5	31.3	40.2	49.2
Sweet potato (*Ipomoea* sp.)	452	1.0	19.7	25.3	31.0
Plantain (*Musa* sp.)	469	0.9	17.1	22.0	26.9
Cassava (*Manihot* sp.)	565	1.0	15.8	20.3	24.8
Sagopalm (*Metroxylon* sp.)	1494	1.4	8.4	10.8	13.2

Source: Basal metabolic rate from the Schofield (1985) prediction equation using body weight. Food composition values from the Food Composition Table for use in East Asia (United States Department of Health, Education and Welfare, 1972).

intake), while five would not. Those five are yam, sweet potato, plantain, cassava and sagopalm.

The situation is not quite the same with children, however. Protein requirements are highest in early life when growth is rapid, and the weaning diet of some groups is likely to give an inadequate intake of protein to young children. Shortages of other nutrients are also possible. In addition to protein, it has been suggested that zinc (Golden, 1988; Ann Prentice & Bates, 1994), calcium (Fraser, 1988) and possibly other deficiencies may contribute to the growth faltering experienced by children in the developing world.

Therefore, in studies of human adaptation and adaptability, it cannot always be assumed that energy is the primary nutritional stressor, although this is likely to be the case in most situations. As with any research tool, therefore, its usefulness needs to be evaluated and not assumed. Consideration of some early studies of ecological anthropology using energy as an analytical tool may be instructive in this regard.

Early studies incorporating measures of energy

The first anthropologist to consider energy was White, in his examination of the capacity of different human groups to harness increasing amounts of energy (White, 1949, 1959). This was followed by a number of studies in the 1960s using energy as the core currency in ecological exchange. From the perspective of the 1990s, these were rather crude and often flawed, but they generated a body of data and an understanding of traditional subsistence systems which are debated to this day. Notable studies were those of Lee (1965) of the subsistence ecology of the !Kung bushmen, and Rappaport's (1968) study of the Maring ritual cycle, as it related to sweet potato cultivation and pig husbandry.

The work of Richard Lee

Lee had three aims in his examination of the ecological basis of an African hunting and gathering economy. These were (1) to outline the subsistence strategy which allowed the !Kung bushmen to live well in a harsh environment with only rudimentary technology; (2) to show that the !Kung had an elementary form of economic life; and (3) to trace, from a primate baseline, the origin and evolution of human energy relations. In carrying out his study, Lee was careful to exclude all !Kung who were reliant, to any extent, on neighbouring Bantu cattle herders for work and food.

In summary, Lee (1968) demonstrated that the !Kung in the Dobe area where he worked could derive an adequate living from only a modest expenditure of time and effort, challenging the conventional wisdom of the time that hunter–gatherer subsistence was conditioned by scarcity and that life was a constant struggle. That the !Kung had an elementary economy was illustrated by Lee in his descriptions of generalised reciprocity, in which sharing of resources was universal, and the accumulation of surplus non-existent. Lee's study was important, therefore, because it shed new light on hunter–gatherer subsistence.

Problems arose, however, when Lee's conclusions were taken to represent all hunter–gatherer groups. The !Kung are not 'typical' hunter–gatherers. Indeed, no such thing exists, as is amply demonstrated in the benchmark volume on hunter–gatherers edited by Lee and de Vore (1968). It is, therefore, impossible to generalise from one group to all others, past and present. Lee's work has been criticised for a lack of representativeness in a number of areas. Although true hunter–gatherers, the groups he chose were not typical of all contemporary !Kung. However, Lee made this clear when reporting his observations. Furthermore, the study period did not take into account seasonality of food availability (Wilmsen, 1978), and his results are unlikely to generalise across the entire year for the !Kung alone.

The !Kung have been shown to exhibit signs of energy nutritional stress (Truswell & Hansen, 1976) and have a high infant-mortality rate, which increases across the parity of the mother (Pennington & Harpending, 1988). This would suggest that they are not living in a state of 'primitive affluence'; rather, their population is regulated by a number of constraints. It has been suggested that their fertility may be regulated by energetic stress of one sort or another (Bentley, 1985) and that their low work output, in terms of the proportion of the population engaged in hunting or gathering at any time, may be a way of regulating the use of resources in relation to their population size (Ulijaszek, 1993a). That is, the !Kung population may be close to carrying capacity.

The challenging view that Lee (1969) put forward has been resisted on the grounds of poor generalisability. Furthermore, his conclusions have been questioned after more detailed investigation and analysis of different aspects of the energetics of !Kung life, notably energy nutritional status, energy balance in pregnant and lactating women and the seasonality of available sources of dietary energy.

The work of Roy Rappaport

Rappaport carried out fieldwork among the Tsembaga Maring, a group of swidden cultivators in Papua New Guinea (PNG) in 1962 and 1963. He applied ecosystemic methodology in attempting to understand how ritual acts as a mechanism that regulates some of the relationships of the Tsembaga with components of their environment. The Tsembaga Maring are sweet-potato cultivators who practise pig husbandry. They engage in cycles of warfare and pig raising which are punctuated by ritual pig-kills involving large numbers of animals and exchange within and between warring groups. Rappaport used an ecological energetics approach to explore the rationality of the Tsembaga system.

In particular, he examined the part that the pig-killing ritual, which was carried out to appease the ancestors, played in the following: (1) the relationships between people, pigs and gardens, and competition of different human groups for limited land resources; (2) the regulation of the slaughter, distribution and consumption of pigs and its relationship to dietary protein requirements; (3) the regulation of the consumption of non-domesticated animals; (4) the redistribution of the population over time across available land, and between territorial groups; and (5) the regulation of the frequency of warfare and the severity of intergroup fighting. A summary of the ritual cycle is given in Fig. 1.3.

The length of the ritual cycle is largely regulated by the demographic fortunes of the pig population. Rappaport (1968) calculated that on

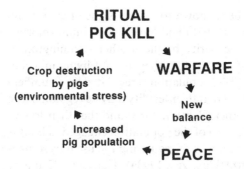

Figure 1.3.   The Tsembaga Maring (Papua New Guinea) ritual cycle.

average a cycle lasts between 12 and 15 years. The ritual signals the time to resume hostilities with a neighbouring group; this is the time when territorial boundaries are removed. Warfare proceeds for a number of years and ceases when a new balance between the population of one group and the neighbours is established. Then follows a period of peace, when the pig population, which is fed cultivated yams and sweet potatoes, increases. The time taken to grow enough pigs to appease the ancestors varies according to how 'good' or productive the land is, but it is at least 6 years. Eventually, the pig population is unmanageable; the pigs do not get enough food from cultivated sources, so they take to foraging and begin to destroy gardens and consume tubers planted for human consumption. At this point, a festival is held in which the vast majority of pigs are slaughtered, reducing the amount of land needed to sustain the combined populations of pigs and human beings. Hostilities are resumed with neighbouring groups, the environmental stress of maintaining the pigs signalling the time for a sacrifice, and the sacrifice signalling the time to attempt to obtain more resources by extending the land boundaries through warfare.

Whether or not warfare resulted in increased land availability then depended upon the numerical strength of the neighbours who were being opposed at that time, and the new boundaries, Rappaport claimed, represented the new balance in land according to population size. In this way, there was an ecological regulation of the human population to land availability, mediated by the pig population growth and slaughter.

Rappaport claimed that pigs, ceremony and nutritional stress all played their part in ecosystemic regulation for the Tsembaga. The gardening of sweet potatoes generated surpluses of dietary energy which was used to feed domesticated pigs, who were a poor source of energy but a good source of protein. Pigs and wild animals provided only small amounts of protein

Table 1.2. *The price of pigs in Papua New Guinea (time devoted to pig husbandry and root-crop cultivation by a group of 204 Tsembaga Maring)*

	Total work input	
	person hours[a]	(%)
Root cultivation		
For human consumption	33920	54
For pig consumption	10710	17
Herding pigs	17000	27
Hunting and gathering	1340	2
Total	62970	

Source: From Rappaport (1968).
[a] Total person hours measured across one year 1962–3.

on a day-to-day basis, and Rappaport found some evidence of protein deficiency in the Tsembaga. He argued that the practice of killing and consuming pigs at times of misfortune and emergency provided 'physiological reinforcement' when it was needed to those that needed it, and was, thus, an effective way of using the limited amounts of animal protein available. Energy from the highly successful sweet potato staple was the key resource in fuelling the ritual cycle. Rappaport's use of energetics in analysis of Tsembaga subsistence has provided some understanding of the way in which the cultivation of a high-yielding, energy-rich staple crop, such as sweet potato, gives a high dietary energy return for the amount of human labour required to cultivate it and generates energy surpluses which can be used to generate energy-poor, but protein-rich resources, such as pigs (Table 1.2), which contributed only 8% of total energy intake, compared with 91% from root crops (Table 1.3).

A criticism of his original thesis, however, is that times of nutritional stress did not usually coincide with the cycle of ritual pig slaughters that were used to appease the ancestors. Therefore, it is difficult to claim that human population pressure was the signal for pigs to be killed and warfare resumed. This is supported by analyses from other investigators that the Tsembaga are rarely anywhere near their carrying capacity. Furthermore, although the intakes of protein in Papua New Guinea are the lowest in the world (Koishi, 1990), the energy subsidy used in pig raising does not appear

Table 1.3. *Daily energy intake of Tsembaga,*
by source

	Daily energy intake (% of total daily intake)
Source	
Root crops	91
Pigs	8
Hunting and gathering	1
Total daily intake	
Adult males	10.77 MJ/day
Adult females	9.05 MJ/day

Source: From Rappaport (1968).

to provide a boost in protein intake at an appropriate time for the Tsembaga, as Rappaport claimed.

In a critique of Rappaport's work, McArthur (1977) determined that over the 5 days of pork consumption during the ritual kill that he observed, 1 kg per day would be eaten, which would supply over 12 MJ of dietary energy and 109 g of protein. For an average 43 kg Tsembaga male (Rappaport, 1968), a safe level of protein intake would be about 32 g per day (FAO/WHO/UNU, 1985). Such high levels of pork consumption over a period of only a few days is an inefficient use of protein, since, taking the case of the adult male consuming 1 kg of meat per day, the body would use at the very most only a third of this protein in protein anabolic processes, while the rest would be deaminated and used as an expensive source of metabolisable energy. It is possible that such high intakes of pork over short periods of time may actually be harmful. High intakes of protein in subjects habitually consuming very small amounts of protein have been shown to be associated with the enteropathy pig-bel, or enteritis necroticans, which in its most severe manifestation leads to death (Lawrence, 1992).

Energetics and anthropology
Patterns of energy production and use in a community or population are a function of the type and extent of local energy acquisition, exchange relationships with other groups and other activities that are important for the maintenance of group function, including individual biological needs as well as group economic maintenance and reproduction (Nydon & Thomas, 1989). These are summarised in the flow diagram (Fig. 1.4).

Dietary energy generated through various processes involving energy

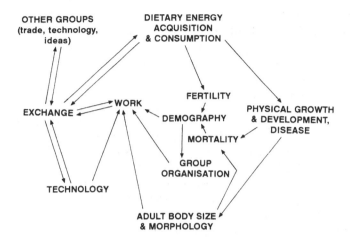

Figure 1.4. Human ecological relationships.

expenditure can be divided and consumed in a variety of ways, which, in turn, influences the pattern of energy expenditure, storage and mobilisation of bodily energy stores. The relationship between intake, expenditure and storage within group or community is usually complex but can be studied using appropriate energy balance (Ulijaszek, 1992a) and energy accounting methods. In the studies carried out by Lee (1969) and Rappaport (1968), energy was shown to be important in the regulation of human activity, although not as universally as the authors might have hoped. Studies carried out in the 1970s were more sophisticated in their methodology and cautious in their interpretations than those carried out by Lee and Rappaport, respectively. In particular, the relationships between physiological processes associated with energy expenditure and energy balance in the individual and energy transfer processes at community level have been examined.

Thomas (1973a, 1976) described the energetics of a high Andes Quechua community in the Nunoa District of Peru. He highlighted the importance of trade in animal products for the purchase of wheat flour and other imported foods, with groups living at lower altitudes, for the nutritional maintenance of their high population levels, relative to natural resources. In effect, the Quechua in Nunoa District have reached the carrying capacity of their land and rely on the husbanding and sale of the highly prized commodities, meat, fur and leather for their survival. In addition, he concluded that the Quechua could not sustain themselves without high levels of child labour in the herding of animals, freeing adults to carry out labour-intensive agricultural tasks. The mean completed fertility rate in the 1960s was 6.7 children (Hoff, 1968). With this high fertility rate providing a large child labour pool, demographic stability is maintained by migration, which can be seasonal or occur at times of scarcity or hardship and may be either temporary or permanent.

Other studies have been carried out in semi-industrial systems, such as that of the people inhabiting the Polynesian atoll of Ontong Java (Bayliss-Smith, 1977), where energetics methods were used to examine the impact of industrial technology on a small-scale society. Bayliss-Smith showed that, despite increased inputs of energy from fossil fuels into subsistence work and greater productivity and export of surpluses, the overall efficiency of the system is no greater than it would be were there no subsidy from fossil fuels. Therefore, the increased export of products resulting from greater inputs largely benefits the world market, and not the producers. However, surplus income was available for the purchase of imported manufactured goods, something held as important by many modernising societies in the Pacific region and elsewhere.

Another semi-industrial system was studied in 1955 by Epstein (1962) in Tamil Nadu, South India. Observations repeated in 1975–6 (Rebello *et al.*, 1976) allowed a comparison of the energetics of rice farming over a period in which the impact of green revolution inputs on group labour activity and productivity could be assessed (Bayliss-Smith, 1982a,b). Overall, total food energy production increased by 57%, with 111% increase in energy inputs, mostly from fertiliser and fossil fuels. Thus, the aim of green revolution inputs, which were to increase yields per unit area of land, have succeeded in this region, but only at the cost of energy subsidies which exceed the increased energy output. The comparison showed that the efficiency of production had declined.

By the mid 1980s, a number of studies of systemic ecological energetics had been carried out in a wide variety of contexts. In addition, data collected as part of earlier studies, prior to the onset of the ecological energetics movement, had been reanalysed in energetics terms. Details of most of the studies carried out up to that time are given in Table 1.4. This body of data has allowed a comparison of the energetic efficiencies of different subsistence systems operating at different levels of technology.

A number of developments in the fields of human physiology and nutrition also began to be incorporated into anthropological inquiry in the 1980s. These included the changing understanding of the energetics of bodily maintenance, locomotion, work and physical work capacity, pregnancy, lactation and physical growth, and the methods for measuring it (Ulijaszek, 1992a). Therefore, it became possible to examine the energetics of an ecosystem in more detail by considering such things as the efficiency of work and the energy expended therein, the rationale of working in groups or bands as opposed to individually, and the impact of undernutrition on the ability to perform subsistence tasks.

Another development was the use of predictive models in examining variations on existing subsistence strategies, or the effects of change on different measures of ecological success. This was made possible by the availability of computing power which could perform the sometimes complex and usually tedious calculations involved in modelling. Notable among these were the analyses of !Kung female foraging strategies (Blurton Jones & Sibly, 1978), the impact of different types of change on highland Quechua populations (Thomas *et al.*, 1982) and the rationality of post-harvest gorging by Gambian agriculturalists (Dugdale & Payne, 1986). Modelling is useful in hypothesis generation and in understanding complex interactions. Drawbacks include a limited ability to predict outcomes in any realistic manner, and poor generalisability to groups other than those from which the data used in model building were obtained.

Improved field methods and modelling techniques have led to several theoretical advances, including: (1) theories of hunter–gatherer foraging strategies based on optimality modelling; (2) a changed understanding of reproductive ecology resulting from advances in the measurement of reproductive performance and the energetics of pregnancy; and (3) an improved understanding of the importance of work and work organisation, based on improved methods for the observation of physical activity and a changing knowledge of the energetics of human effort.

Energetics has entered the study of reproductive ecology (Rosetta, 1990, 1994; Ellison, 1991) since many of the biological stresses influencing reproductive function are energetic (Lunn, 1994). There have been reductions in total fertility rates in most parts of the developing world, but especially in countries experiencing some level of economic growth (United Nations Children's Fund, 1990). Smaller family size has implications for dietary energy requirements across the lifecycle of the family and for subsistence productivity. Total food requirements may be lower for a smaller family, but the potential family work-force is also smaller. The question of what constitutes reproductive success in humans becomes more complex and more difficult to answer as populations become more land-limited and integrated into the global cash economy. Generally, proxies such as nutritional or health status are taken as yardsticks of such success, but studies of energetics, ovulatory function, pregnancy, lactation and physical growth are beginning to create a new understanding of these

Table 1.4. *Populations in which studies of ecological energetics have been carried out*

Group or location	Country	Subsistence pattern	Study period	Reference
Jiaxing	China	Rice cultivation	1765	Chen Hengli, 1958 Calcualtions: Dazhong & Pimentel, 1986
Wiltshire	United Kingdom	Wheat and barley cultivation, sheep farming	1820s	Cobbett, 1830 Calculations: Bayliss-Smith, 1982a,b
Iban	Sarawak, East Malaysia	Rice cultivation	1949–51	Freeman, 1970 Calculations: Harris, 1971
Genieri	Gambia	Cereal and peanut cultivation	1947–9	Haswell, 1953 Calculations: Harris, 1971
Karnataka	India	Rice cultivation	1955	Epstein, 1962 Calculations: Bayliss-Smith, 1982a,b
Hanunoo	Philippines	Rice cultivation	1957	Conklin, 1957
Tepozoztlan	Mexico	Maize cultivation		Lewis, 1963 Calculations: Weiner, 1972
Hadza	Tanzania	Gathering, hunting	1958–60	Calculations: Pimentel & Pimentel, 1979; Harris, 1971 Woodburn, 1972
Lamotrek Atoll	Micronesia	Root cultivation, fishing	1962–3	Alkire, 1965 Calculations: Odum, 1971
Tsembaga Maring	PNG	Root cultivation, pig husbandry	1963	Rappaport, 1968
!Kung San	Namibia	Gathering, hunting	1963–5	Lee, 1965
Bomagai Angoiong Maring	PNG	Root cultivation, pig husbandry, hunting	1964–5	Clark, 1971
Moscow Oblast	Russia	Collective farm: wheat, potatoes or fodder crops, barley or oats, and grass with clover, in rotation	1966 (statistical average)	Moscovskaia Oblast, 1967 Calculations: Bayliss-Smith, 1982a,b
Raiapu Enga	PNG	Root cultivation, pig husbandry	1966–7	Waddell, 1972 Calculations: Morren, 1977; Bayliss-Smith, 1977

Group	Location	Subsistence	Year	References
Quechua, Nunoa	Peru	Herding, root cultivation	1967	Thomas, 1973b
Eskimo	Baffin Island, Canada	Hunting, fishing	1967–8	Kemp, 1971
Tashpauni Miskito	Nicaragua	Root and rice cultivation, fishing, hunting	1968–9	Nietschmann, 1973
Miyanmin	PNG	Root cultivation, pig husbandry, hunting	1969	Morren, 1977; Little & Morren, 1977
Ruhua Nuaulu	Seram, Indonesia	Root cultivation, gathering, hunting	1969–71	Ellen, 1978
Ontong Java	Solomon Islands	Root and coconut cultivation, fishing	1970–71	Bayliss-Smith, 1977
Wiltshire	United Kingdom	Large farm: barley and wheat cultivation, sheep and cattle husbandry	1971 (statistical average)	Ministry of Agriculture, 1973; Leach, 1976 Calculations: Bayliss-Smith, 1982a,b
Wonie	PNG	Sago horticulture, hunting, fishing	1971–2	Ohtsuka, 1983
Nacamaki	Koro, Fiji	Root and coconut cultivation, fishing	1973–4	Bayliss-Smith, 1977
Nasaqalau	Lakeba, Fiji	Root and coconut cultivation, fishing	1974	Bayliss-Smith, 1977
Alyawara	Australia	Hunting and gathering	1974–5	O'Connell & Hawkes, 1981
Karnataka	India	Green revolution, rice cultivation	1975	Rebello et al., 1976 Calculations: Bayliss-Smith, 1982a,b
Wopkaimin	PNG	Taro horticulture, hunting	1975	Hyndman, 1979
Etolo	PNG	Hunting and horticulture, sago	1979–80	Dwyer, 1983
Arnhemland	Australia	Hunting and gathering	1979–80	Altman, 1984
Ache	Paraguay	Hunting and gathering	1981–2	Hawkes et al., 1982
Maharashtra	India	Rice and sorghum cultivation	1983	Edmundson & Edmundson, 1989

ecological relationships. However, there are a number of methodological problems that have yet to be resolved including: (1) the use of data collected using a variety of techniques, including biological and social variables, without undue loss of accuracy and precision; (2) the development of modelling and statistical techniques that will allow complex relationships to be understood; and (3) the development of longitudinal data collection and modelling techniques that allow the effects of change to be observed and analysed.

The well-being of the mother and the children she bears are closely related. Dugdale (1986) has suggested that the mother and nursing infant should be considered as a unit, or dyad. The size of the infant is related to the maternal size and the environment which she experiences, the breastmilk she produces in response to the child's demands and other feeding behaviours which she introduces to the child. New methods of, and increased accuracy in, energy-balance measurement have led to greater understanding of the energetics of child-bearing. In most traditional societies, the pattern of human growth is different from that of populations in industrialised nations, resulting in smaller adult body size. The suggestion that smaller body size of adults carries no cost to biological function (Seckler, 1982) has been convincingly discredited (Satyanarayana *et al.*, 1977; Spurr, Maksud & Barac-Nieto, 1977; Brooks, Latham & Crompton, 1979; Immink & Viteri, 1981; Immink *et al.*, 1984; Spurr, 1988a; Ferro-Luzzi *et al.*, 1992; Strickland & Ulijaszek, 1993), while the process of becoming small, or growth-retarded relative to western growth references, is associated with greater risk of death in most populations (Pelletier, 1991).

The energetics of work must be considered in relation to subsistence practices and behaviour. The organisation of work by individuals and groups varies with the types of subsistence practice adopted. The formation of communities, villages or settlements, although fundamentally social, has consequences for energy balance. For example, social relations may involve reciprocity in food distribution and communality in work-group formation. Both acts can serve to reduce the level of energetic stress experienced by individuals within the groups. Ecological anthropologists are often concerned with the energetic rationale of different types of subsistence strategy, either in the context of other activities or when attempting to plot the evolution of such strategies in relation to land availability and demographic pressure.

In seasonal environments, the requirement for hard work varies across the year, as does the availability of food, often resulting in fluctuating energy balance across the year (Ferro-Luzzi & Branca, 1993). It is possible that food sharing and work-group formation may not successfully buffer all individuals under such conditions. The identification of individuals or groups at risk of energy stress is useful in attempting to understand the process of adaptation in any population or community, as well as being of potential public health significance.

Finally, a new approach in evolutionary ecology has involved energetics modelling. Data on the energetics of locomotion and of growth and development in contemporary human and non-human primate populations has been used in association with body size estimates for extinct hominids to examine the energy cost of encephalisation (Foley & Lee, 1991; Leonard & Robertson, 1992, 1994), increased body size (Leonard & Robertson, 1994) and bipedalism (Leonard & Robertson, 1992).

Summary

In this chapter, the use of ecological energetics and energetics in human adaptability is described. The history of ecological energetics is a brief one, beginning in the 1960s. Early studies were crude but provided a new quantitative approach to the study of traditional subsistence economies, using systems theory. Since then, estimates of energy intake, expenditure, cost of activity, balance and flow have been used in attempts to understand human subsistence within the adaptation and adaptability framework. Research has focussed on: (1) how the need for dietary energy and the ways in which it is obtained affect different aspects of human population biology or ecology; (2) the implications of different subsistence and foraging strategies for fertility and biological fitness; and (3) human responses and adaptations to seasonal energetic stresses. Modelling procedures have allowed energetics data to be used predictively and in hypothesis generation, in these areas and in some aspects of evolutionary ecology.

Energy is an interconvertable currency that can be used in quantitative analysis of activities of human groups that are minimally or only partially involved in the cash economy. Thus, although the approach may not lead to a comprehensive understanding of human adaptation, it can give an extensive account of that aspect of human functioning related to resource acquisition, subsistence and ecological and reproductive success. Furthermore, improved methodology and precision of measurement of different components of individual energy expenditure have allowed anthropologists to address such issues as: (1) the energetics of different physiological states, including undernutrition, obesity, pregnancy and lactation; (2) the ecological correlates and functional consequences of small body size; (3) the influence of energetic stress on ovulatory function; and (4) the levels of work effort and output in different types of subsistence system.

Theory and methods

2 *The individual and the group*

Energetics allows two fundamental aspects of human existence to be explored: production and reproduction. Differential reproductive success is a direct measure of fitness in a population, while the ability to acquire resources, food in particular, through productive methods is an indirect measure of fitness, since it can influence reproductive success.

Although a number of energetics measures can be made at group or individual level (Chapter 3), aggregating measurements made on individuals allows variation within a population to be examined. Since differentials in reproductive performance and resource acquisition can lead to differences in reproductive success, such within-population variation is of interest to biological anthropologists. Variation is known to exist for all components of energy expenditure, including: (1) total daily energy expenditure (TEE); (2) basal metabolic rate (BMR); (3) energy expended in physical activity; (4) physical work capacity (PWC); (5) the energy costs of pregnancy and lactation; (6) the thermic effect of food (TEF); and (7) the energy costs of physical growth. Although most of these measures are related to body size, in many instances some variation remains, even after correcting for interindividual differences in body weight. In this chapter, the components of energy expenditure at the individual level are examined and related to ecological processes at group and population level.

Energy balance

Humans, like all mammals, are ruled by the laws of thermodynamics, the law of conservation of energy being the one which underpins bioenergetics as a discipline (Kleiber, 1961). Briefly, the law states that energy can neither be gained nor lost but may be transformed in nature. In living systems, this includes the transfer of chemical energy into heat, work and electrical energy, and the interconversion of chemical energy from one type to another. This is also one aspect of the study of physiological nutrition, and it cannot be ignored that chemical energy, in the form of food, can also be used to create morphological structures during growth and development. Furthermore, these structures incorporate macronutrients, predominantly protein but also lipids, and are primarily functional; breakdown of these

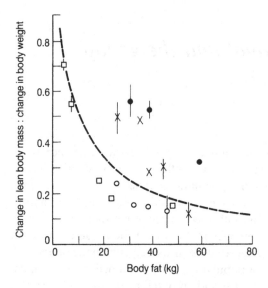

Figure 2.1.   The proportion of body weight lost as lean and fat tissue at different levels of body fatness. The curve is drawn from results of weight-loss experiments on adult subjects, involving energy intakes of not less than 4.2 MJ/day. Energy intake (kcal/day): 0–440 ($n = 25$) (●); 500–900 ($n = 77$) (X); 1000–1400 ($n = 81$) (○) 1500–1900 ($n = 80$) (□). (From Ferro-Luzzi & Branca, 1993; modified from Forbes, 1989.)

structures to liberate macronutrients to meet energy needs can in some circumstances be to the detriment of physiological function. The energy balance equation is simply written:

Energy balance = Energy in − Energy out

If output exceeds input, there is negative energy balance, and body mass will be mobilised to meet energy needs. If prolonged, negative energy balance will result in weight loss, the composition of which will depend on body composition at the onset of imbalance: someone with greater body fatness will lose more fat in relation to muscle than someone with lower fatness. Figure 2.1 illustrates the relationship between the proportion of body weight lost as lean tissue at different levels of fat stores (Forbes, 1989; Ferro-Luzzi & Branca, 1993). In general, individuals in industrialised countries lose a greater proportion of body weight as fat than do subjects in the rural developing world, because they start with a higher proportion of body weight as fat. For example, Scottish males between the ages of 40 and 70 years have 25% of their body weight as fat (Durnin & Womersley, 1974)

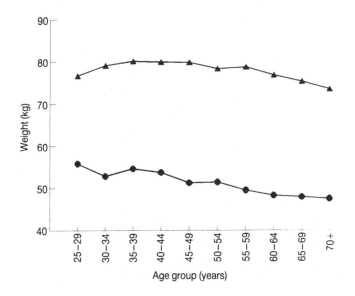

Figure 2.2. Mean body weight by age group for adult male Iban, Sarawak (●) and European Americans (▲). (From Strickland & Ulijaszek (1993); Frisancho (1990).)

and under conditions of weight loss are predicted to lose about 70% of total weight as fat. Conversely, adult males in Papua New Guinea are among the leanest in the world, with about 10% body fat (Norgan, 1982), and are predicted to lose about 50% of total weight as fat, under the same conditions.

If intake exceeds expenditure, the resulting positive energy balance, if prolonged, will result in weight gain. Three patterns of energy balance are illustrated in Figs. 2.2 and 2.3. Figure 2.2 shows the mean weight gains and losses across the course of a lifetime typical of adult males in industrialised and developing countries, respectively. The Iban of Sarawak are upland rice cultivators operating at very low levels of technology (Strickland & Ulijaszek, 1993), while the European Americans are from the NHANES III study (Frisancho, 1990). Between the ages of 35 and 40 years, Iban males are lighter than subjects between the ages of 25 and 30 years. Conversely, American males become heavier across the same period. Once above the age of 60 years, Iban males are about 15% lighter than their counterparts in early adulthood, while North American men are at a similar weight. These weight changes across life are to be observed in many populations and are the result of small overall energy imbalances across prolonged periods.

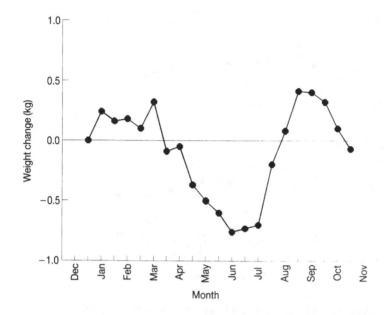

Figure 2.3.   Body weight changes across 1986 for a group of adult women in Benin. Weight change is expressed as differences from the baseline body weight in December 1985. (From Schultinck *et al.*, 1990.)

In developing countries, particularly among rural populations experiencing seasonality of one or several types (Ulijaszek, 1993b), there is movement from positive to negative energy balance across the year (Fig. 2.3). Rural Beninese women show a fairly typical pattern of body-weight change across the agricultural calendar as a result of these imbalances (Schultinck *et al.*, 1990), weight loss taking place during the rainy season, weight gain after the maize harvest. Therefore, approximate energy balance may only be achieved across the course of an entire subsistence cycle, usually a year (Ulijaszek & Strickland, 1993a). In many cases, it is likely that this balance is not absolute, a small imbalance across the year leading to overall weight loss across the adult lifespan.

Components of energy intake and expenditure

The existence of intermediary metabolism enables humans and other animal species to use carbohydrate, fat and protein interchangeably as dietary energy. The subject of macronutrient metabolism is properly that of nutrition but is of relevance to bioenergetics in as much as the patterns of ingestion and use for energy expenditure carry with them varying degrees of metabolic energetic efficiency (Flatt, 1985). The carbohydrate component of body composition is glycogen and is extremely small relative to body muscle and protein mass and fat mass.

Human diets vary in their relative composition of carbohydrate, fat and protein. In general, however, carbohydrate forms the largest component of intake in all societies, except strict hunter–gatherers (Table 2.1). If intake of this macronutrient exceeds requirements for immediate energy expenditure, then the surplus is stored predominantly as fat. Fat intake varies enormously between traditional societies, with hunter–gatherer groups having intakes, when expressed as a proportion of total daily energy intake, similar to, or in excess of, those found in industrialised populations. In Britain, total daily energy intake from fat is around 40% (Nelson, 1993). If intake is in excess of the immediate requirements for free energy, the surplus is stored predominantly as fat. This is also the case if protein intake exceeds maintenance requirements for protein and energy. Central to these processes are the metabolic pathways of glycogenesis and glycogenolysis (formation and breakdown of glycogen), glycolysis (breakdown of glucose, the common intermediary of carbohydrate metabolism), beta-oxidation of fatty acids (the common intermediaries of fat metabolism), fatty acid synthesis, and transamination and deamination (allowing interconversion of many of the amino acids, and making the carbon skeletons of amino acids available for energy metabolism). The tricarboxylic cycle links these pathways, both replenishing anabolic pathways when required and providing reducing equivalents for the regeneration of ATP.

Under conditions of negative energy balance, glycogen, fat and protein are used as metabolic fuel in the absence of dietary energy. The relative efficiency of oxidation of these endogenous substrates varies. The complete oxidation of bodily fat to carbon dioxide regenerates ATP with 94% efficiency, liver and muscle glycogen with 88% and 19% efficiency, respectively, and muscle-glycogen with 88% and 19% efficiency, respectively, and muscle-bound protein with 79% efficiency (Flatt, 1985). Since leaner subjects lose a greater proportion of lean body mass relative to fat mass during weight loss than do less lean individuals (Forbes, 1989), the composition of weight loss will directly influence the efficiency of use of bodily tissues for free energy; individuals and populations with lower proportions of body fat at the time of onset of negative energy balance will metabolise their body mass less efficiently than those subjects and groups with higher proportions of their body weight as fat.

The process of oxidative phosphorylation takes place in the inner mitochondrial membrane and is the stage at which the common energy currency of ATP is regenerated from ADP. The process is driven by the

Table 2.1. *Macronutrient composition of selected human diets*

Country	Group	Subsistence base	Percentage of total daily energy intake from:			Reference
			Carbohydrate	Fat	Protein	
PNG	Ningerum	Horticulture	91	5	4	1
PNG	Lufa	Horticulture	82	10	8	2
Miyanmar	Rural	Agriculture	77	13	10	3
Cameroon	Yassa	Fishing and agriculture	77	11	12	4
Cameroon	Bakola	Foraging and horticulture	77	10	13	4
Nepal	Rural	Agriculture	76	13	11	5
PNG	Kaul	Horticulture	75	18	7	2
Afghanistan	Nurzay Pashtun	Agro-pastoralist	75	13	12	6
Kenya	Kikuyu	Agro-pastoralist	72	9	19	7
PNG	Gidra	Horticulture	70	25	5	8
Barbados	Rural	Agriculture	69	23	8	7
Cameroon	Mvae	Hunting and agriculture	65	21	14	4
Namibia	!Kung	Hunter–gatherer	21	62	17	9
Canada	Eskimo	Hunter–gatherer	7	45	48	7

Sources: 1, Ulijaszek, 1992b; 2, Norgan et al., 1974; 3, Tin-May-Than & Ba-Aye, 1985; 4, Koppert et al., 1993; 5, Brown Worth & Shah, 1968; 6, Casimir, 1991; 7, in Harrison et al., 1964; 8, Ohtsuka, 1990; 9, Lee, 1968.

active transport of protons from within the inner mitochondrial membrane and the phosphorylation of ADP by the enzyme ATP synthase in the presence of re-entering protons (Mayes, 1988). However, the permeability of this membrane is influenced by hormonal regulators of energy balance, notably thyroid hormones (Dauncey, 1990), and hence the energetic efficiency of ATP production from given amounts of the reducing equivalents $NADH^+$ and $FADH_2$ generated in the course of intermediary metabolism is also influenced.

Mitochondria provide the sites of ATP regeneration for use in processes involving energy expenditure; they are also extremely responsive to changes in energy balance, energy flux and macronutrient content of the diet (Aw & Jones, 1989). In particular, increases in thyroid hormone levels in response to changing, positive energy balance have been associated with increases in the area of the mitochondrial inner membrane and increased amounts of respiratory chain proteins (Dauncey, 1990). This is a response to increased requirements for energy expenditure. At the other extreme, deficiencies of protein (Svobada & Higginson, 1964) and starvation (Rohr & Riede, 1973) have resulted in mitochondrial giantism in animal models, with attendant degenerative changes in cristae structure and the development of a matrix which is translucent to electrons (Tandler & Hoppel, 1986). An important consequence of this is decreased substrate oxidation (Tandler & Hoppel, 1986). Mitochondrial giantism is probably a pathological state, to be found only in individuals beyond the limits of human energetic adaptability. However, there is evidence that a less extreme macronutrient deficiency, that of protein, can alter mitochondrial structure in one species of non-human primate *Macaca mulatta* (Ordy et al., 1966).

Other factors that can affect mitochondrial number and plasticity include hypoxia and endurance training. This is of relevance to human ecological energetics, since this type of energetic adaptability could take place in human populations living at high altitude, or who perform subsistance tasks which require high levels of physical performance, endurance, or both. During acclimatisation to chronic hypoxia, both mitochondrial number (Ou & Tenney, 1970) and volume (Boutellier et al., 1983) are increased, and there is an increase in the activities of mitochondrial enzymes, notably the cytochrome oxidases (Bouverot, 1985). These changes reflect the increased utilisation of protein and fat as energy substrates, in place of glucose (Blume, 1983). The number and size of mitochondria also change in response to submaximal endurance exercise training, the increase being limited to muscles that participate in training (Holloszy & Coyle, 1984). There are also increases in tricarboxylic acid cycle enzymes, pyruvate and fatty acid oxidation, electron transport

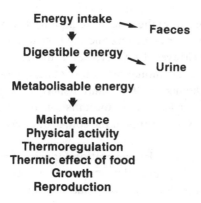

Figure 2.4.   The fate of dietary energy in the body. (From Ulijaszek, 1992a.)

activity (Terjung & Kaciuba-Uscilko, 1986; Williams, 1986), and a shift in substrate utilisation from carbohydrate to fat (Ahlborg *et al.*, 1974) that serves to spare glucose, allowing a greater efficiency of energy-utilising processes with less glucose depletion (Aw & Jones, 1989). In both hypoxia and endurance training mitochondrial adaptability provides for improved energetic efficiency.

The fate of dietary energy entering the body is given in Fig. 2.4. Indigestible energy is voided in the faeces. Many complex carbohydrates, such as those found in vegetables, are to a great extent indigestible to humans. Although it is generally assumed that the contribution of so-called unavailable carbohydrate to metabolisable energy is small (Paul & Southgate, 1978), the absorption of short-chain fatty acids produced from their bacterial degradation in the large intestine was long ago demonstrated (Grove, Ohmstead & Koenig, 1929) and considered to make a contribution to dietary energy availability (McCance & Lawrence, 1929). Livesey (1990) has shown that the contribution of unavailable carbohydrate to dietary energy intake is about 8.4 kJ/g consumed. Energy losses in the urine are very small, and the metabolisable energy is the energy remaining. This can be used in a variety of ways.

Human energy expenditure can be broken down into a number of components, including the energy cost of maintenance, thermic effect of food, physical activity, thermoregulation, growth and reproduction. These components vary between individuals and populations and differ according to the extent to which they can be modified by human behaviour. The energy cost of bodily maintenance includes all functions that preserve bodily integrity, including cardiac activity, sodium and calcium pumping, sympathetic nervous system activity and homeostatic metabolic processes, including thermoregulation (Ulijaszek, 1992a). An approximation of this can be obtained by the measurement of BMR, or resting metabolic rate (RMR). The energy cost of digestion, absorption, transport, metabolism and storage of ingested food is the thermic effect of food (TEF) and accounts for approximately 10% of energy intake (Horton, 1983). It can only be estimated under controlled laboratory conditions and is rarely measured in the field, although it has been shown that measures similar to TEF, dietary-induced thermogenesis and post-prandial thermogenesis, vary between 6 and 8%, and 5 and 6% of energy intake across seasons for Gambian men (Minghelli *et al.*, 1991) and women (Frigerio *et al.*, 1992), respectively. The TEF is influenced by the macronutrient composition of the diet and the level of energy intake (Belko, Barbieri & Wong, 1986; Kinabo & Durnin, 1990; Norgan, 1990). Energy is expended in physical activity by all individuals at all times, while growth and reproduction carry energetic costs in childhood and in most adult women at some time, respectively.

Factors influencing energy expenditure

Body size and composition are the most important factors influencing maintenance metabolism. The effect of body size on between-population differences in BMR is illustrated in Figs. 2.5 and 2.6, where the BMRs of some groups of adult males in developing and industrialised nations are contrasted. The range of differences in mean values for absolute BMR between non-industrialised and industrialised populations observed in Fig. 2.5 disappears when the values are expressed per kg body weight (Fig. 2.6).

Although variation in BMR becomes smaller when body weight is taken into consideration, lower than expected values still exist, and a variety of possible reasons for the remaining variation have been suggested. It has been argued that expressing BMR per kg fat-free mass does not remove all the non-metabolically active tissue and body size effects and that most of the variation can be accounted for by differences in body composition (Shetty, 1993). It is impossible to determine the effects of small differences in body composition on BMR with the techniques and technologies currently available, since it would be necessary to measure the size of the highly metabolically active tissues of the liver, brain and kidneys (Table 2.2), as well as the body compartments, which can now be measured with some degree of success (Shephard, 1991).

Keys, Taylor and Grande (1973) estimated that the brain and liver account for 3–5% of total body weight while using up to 40% of resting

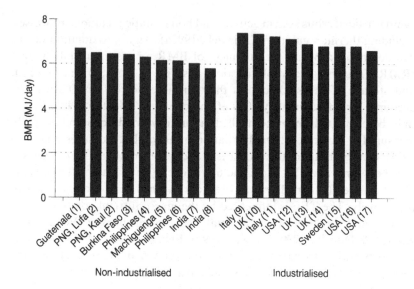

Figure 2.5.   Basal metabolic rate of groups of adult males in
non-industrialised and industrialised countries. Sources: 1, Viteri *et al.* (1971);
de Guzman *et al.* (1974); 2, Norgan *et al.* (1974); 3, Bleiberg *et al.* (1981); 4, de
Guzman *et al.* (1974); 5, Montgomery & Johnson (1977); 6, de Guzman
(1981); 7, Srikantia (1985); 8, McNeill *et al.* (1987); 9, Norgan & Ferro-Luzzi
(1978); 10, Edholm *et al.* (1970); 11, in Quenouille *et al.* (1951); 12, Boothby &
Sandiford (1922); 13, Norgan & Durnin (1980); 14, Ulijaszek & Strickland
(1991); 15, Wising (1934); 16, Benedict *et al.* (1914); 17, Owen *et al.* (1987);
and Clark & Hoffer (1991).

energy expenditure. In addition, Elia (1992) calculated that muscle mass in
an adult male might contribute 40% to total body weight, but only 22% to
BMR. Given these differences in weight and metabolic activity, it is easy to
surmise that the differences in BMR observed between populations and
individuals, after controlling as much as possible for differences in body
size and composition, could be accounted for by small differences in the
ratio of visceral to non-visceral mass.

Differences in energy expenditure owing to physical activity can also be
large. For example, the total energy expenditure of a 75 kg male, whose
BMR predicted from body weight using one of the Schofield (1985)
equations is 7.6 MJ/day, would be 10.7 MJ/day at a physical activity level
(PAL) of 1.4 × BMR, and 22.9 MJ/day at a PAL of 3.0. Thus, the difference
in total energy expenditure between extreme inactivity and extreme activity
is more than two-fold. By far the greatest part of this results from
differences in absolute levels of physical activity, and this is made clear by
the great variation in the energy cost of different physical activities,

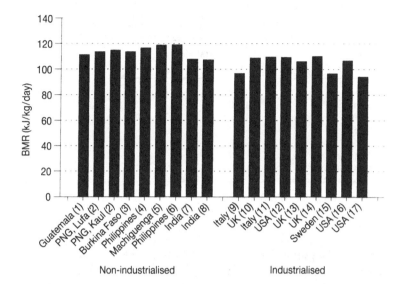

Figure 2.6. Basal metabolic rate of groups of adult males in non-industrialised and industrialised countries, corrected for body weight (kJ/kg per day). (Same sources as in Fig. 2.5.)

expressed as multiples of BMR (Table 2.3). Among the most energetically costly activities are different types of traditional subsistence practice, which, if performed all day long, eventuate in high total daily energy expenditure.

There is also variation in the intensity and energetic efficiency with which muscular work is performed. Intensity of work may be dependent upon time availability for task completion but may also be limited by physical work capacity, which itself may be limited by body mass and composition. Between-group differences in mechanical efficiency of work have been attributed to possible differences in developmental plasticity, as in a comparison of Gurkha soldiers from Nepal with soldiers in the British Army (Strickland & Ulijaszek, 1990), and energy nutritional status, as in studies carried out in Jamaica (Ashworth, 1968), the Gambia (Minghelli *et al.*, 1990) and South India (Kulkarni & Shetty, 1992). Differences in ergonomic efficiency may also contribute to differences in the energy cost of physical activity, which may be universal but may be particularly important in the chronically energy-deficient state (Shetty, 1993).

Behavioural and cultural factors influencing energy expenditure include everything related to production and reproduction, regardless of context. For example, in industrialised countries, subsistence and production for

Table 2.2. *Metabolic rates of different bodily tissues*

Tissue	Daily metabolic rate (kJ/kg per day)
Heart	1840
Kidneys	1840
Brain	1000
Liver	840
Muscle	50
Adipose	20
Miscellaneous[a]	50

Source: From Elia (1992).
[a] Sum of bone, skin, intestines and glands, obtained by difference.

the majority of the population are often indirect and mediated by human energy-saving technologies. Office workers, who may only be involved in production very indirectly, are largely sedentary for the vast majority of the time. Factory workers, directly involved in production, usually have machines and devices which save on physical labour. As a consequence of increasing mechanisation and sedentarisation of the work force, total energy expenditure of populations in industrialised countries has declined in recent decades. In the developing world, patterns of subsistence, including the technologies used, mediate activity patterns and the concomitant total energy expenditures, regardless of whether a group practices hunting and gathering, agriculture at a variety of levels of technology or pastoralism. In these circumstances, the activities which involve greatest energy expenditure may be tied to the production of dietary energy, and linkages between activity, the ability to perform hard physical work, food productivity and nutritional status have been postulated (Martorell & Arroyave, 1988).

Various aspects of reproductive performance involve energetics or are influenced by it, including: (1) fertility; (2) energy expended in pregnancy and pregnancy outcome (one measure of which is birthweight); (3) lactational performance; and (4) child growth and mortality. Cultural factors influencing reproductive performance across the lifespan include determinants of fertility such as age at marriage, coital frequency, use of contraception and perceptions of ideal family size, which in turn may be related to land holdings, patterns of inheritance, extent and type of modernisation, education levels and aspirations for change. These have to be superimposed on biological factors, which may be energetics-related, such as age at menarche, ovulatory function and nutritional status.

Table 2.3. *Energy cost of different physical activities, as multiples of basal metabolic rate*

Activity	No. of subjects measured	Physical activity ratio[a]
Aerobic dancing		
Low intensity	4	5.91
Medium intensity	4	6.31
High intensity	4	8.21
Archery	4	4.35
Army obstacle course	6	4.95
Bed making	5	4.57
Bowls	1	3.75
Carrying 20–30 kg load		
Jungle patrol	7	3.42
Jungle march	2	5.09
Chopping wood with machete	13	3.52
Cleaning windows	1	2.72
Cleaning stairs	20	3.21
Cleaning and drying	5	5.13
Clearing bush	19	4.36
Climbing mountains at own pace at various altitudes		
6000 ft	2	7.98
15 000 ft	3	6.89
20 000 ft	3	6.64
Climbing stairs		
72 steps/min	8	5.19
92 steps/min	3	6.95
Rate unspecified	22	6.79
Rate unspecified, with light load/basket of clothes	2	7.09
Rate unspecified, carrying 11–16 kg load	4	6.83
Coal getting	10	5.65
Coal mining, average cost of activities	12	5.62
Cooking	15	1.80
Cricket		
Batting	4	6.30
Bowling	4	6.35

Table 2.3. (*cont.*)

Activity	No. of subjects measured	Physical activity ratio[a]
Crocheting	1	1.17
Cycling, on flat ground, at own speed	82	5.50
Dancing		
Eightsome reel	3	5.20
Foxtrot	2	3.57
Waltz	2	4.22
Dressing doll	1	1.42
Driving 15 cwt truck	3	2.89
Drill		
Squad of platoon	8	3.14
Church parade	9	3.38
Parade	9	3.38
Competition	11	5.08
Competition practice	11	4.24
Dusting	4	3.66
Field ridging	5	8.33
Finishing copper bands	2	3.78
Fishing		
Setting down net	7	2.53
Pulling up net	14	3.54
Forging	4	3.22
Gardening	11	4.82
Gauging	4	4.18
Golf	1	4.40
Harvesting manioc	11	3.51
Hewing	18	6.07
Hoeing	3	6.53
Hoeing manioc	18	3.78
Hoisting shell, using pulley	1	3.68
Household chores, unspecified	1	3.65
Ironing, using 3.5 kg iron	5	3.50

Table 2.3. (*cont.*)

Activity	No. of subjects measured	Physical activity ratio[a]
Ironing	1	1.46
Kneeling, scrubbing floor	3	2.68
Knitting	1	1.20
Laboratory work	1	2.11
Laboratory work (general)	5	2.38
Labouring	5	5.39
Laundry, (washing, wringing, rinsing, hanging on the line)	4	3.19
Leg exercises (lying on back lifting 9.5 kg load with legs)	12	1.68
Loading	31	6.13
Lying at rest	229	1.05
Maintaining fishing tackle	7	1.86
Making a dug-out canoe	10	4.77
Marching		
With 15 kg load	3	4.64
With 20 kg load; at ease	5	5.08
With 20 kg load: to attention	4	5.14
On level at 5.5 km/hour	13	4.70
Motor cycling on level ground	3	2.39
Musketry	6	2.36
Office work (general)	3	1.18
Operating foot-driven sewing machine	1	1.42
Operating electric sewing machine	1	1.06
Packing batons de manioc	6	1.77
Paddling canoe		
Low intensity	9	3.32
High intensity	15	4.30
Peeling manioc	8	1.86
Planting groundnuts	4	3.08
Planting manioc	11	3.81

Table 2.3. (*cont.*)

Activity	No. of subjects measured	Physical activity ratio[a]
Playing pool	4	3.02
Playing squash	6	8.50
Playing tennis	7	5.83
Playing football	2	5.99
Playing table tennis	1	3.66
Playing cards	5	1.48
Pounding manioc	26	2.57
Pounding rice	4	5.59
Preparing vegetables	12	1.56
Pulling a rickshaw, with 180 kg load (a typical load)	21	10.40
Rifle cleaning	6	2.25
Running downhill (6% gradient, at 12 km/hour)	10	8.16
Running (speed unspecified)	4	7.61
Setting the table	4	3.37
Sewing by hand	2	1.11
Sieving manioc	11	2.19
Sitting at rest	172	1.19
Sitting working	47	1.44
Sitting activities (cooking, washing)	11	2.13
Sitting typing	5	1.69
Sitting activities (coal mining)	19	1.39
Sitting reading, writing, calculating	5	1.36
Sitting typing: electric typewriter	3	1.35
Skiing (cross-country at 4.5 km/hour, with 25–35 kg army pack)	9	7.88
Sports, assorted (relay race, sprint, rope climbing, vaulting and rest period)	12	2.65

Table 2.3. (*cont.*)

Activity	No. of subjects measured	Physical activity ratio[a]
Standing		
At rest	156	5.96
Working	51	1.71
Washing rickshaw	11	2.40
Activities unspecified	37	1.51
Activities (coal mining)	19	1.72
Scrubbing	3	2.45
Stamping	2	3.72
Stamping floor	1	1.71
Timber cutting	26	5.10
Tool setting	5	3.42
Trimming and stripping bark off a tree, using axe and spade	12	7.99
Turning and finishing	8	3.02
Turning		
Heavy	5	3.40
Light	8	2.96
Walking on level		
Speed unspecified	105	3.88
1–2 km/hour	20	2.23
3–4 km/hour	30	2.87
4–5 km/hour	83	3.77
5–6 km/hour	207	4.17
6–7 km/hour	63	5.47
7–8 km/hour	5	7.45
4–5 km/hour, carrying 8 kg load	6	5.13
Walking, coal mining	19	5.31
Walking and carrying	3	3.98
Walking 6–7 km/hour and carrying 70 kg load	1	7.25
Walking uphill, 4–5 km/hour	20	6.59
Walking carrying 11–16 kg load, speed unspecified	4	4.64
Walking uphill at own pace at various altitudes		
6000 ft	1	8.83
15 000 ft	3	7.67
20 000 ft	3	7.43

Table 2.3. (*cont.*)

Activity	No. of subjects measured	Physical activity ratio[a]
Walking downhill, 8% gradient		
1–2 km/hour	10	1.85
3–4 km/hour	10	2.09
4–5 km/hour	10	2.51
6–7 km/hour	10	3.41
Walking downstairs carrying light load/basket of clothes	2	3.36
Walking downstairs, rate unspecified, with load		
15 kg	2	3.60
30 kg	2	4.60
Walking up and down stairs		
97 steps/min	4	6.30
116 steps/min	5	6.87
Washing, dressing, shaving	64	2.69
Washing towels	1	1.88
Washing dishes	1	1.54
Watching football	5	1.84
Weapon training	12	1.82
Weeding, moderate crop of weeds	3	5.03
Working on car assembly line	13	2.36

Source: From James & Schofield (1990); Pasquet & Koppert (1993).
[a] Energy cost of task per minute, divided by BMR per minute.

Nutritional status and energetics-related ovulatory function may in turn be related to subsistence performance, which is also energetics-related. Therefore, the energetics of production and reproduction may in some circumstances be closely interrelated.

Subsistence performance and productivity

Humans use a variety of techniques to acquire resources from the environment, and the range of techniques used and the way in which they are employed constitute the subsistence strategy of a particular group. Understanding any subsistence strategy requires information at the individual and group level. Where hard physical labour is required for at least some of the time, knowledge of individual physiological energetics such as physical work capacity can provide some indication of whether or not this is a limiting factor in resource acquisition. In a similar way to reproductive performance, within-group variation in the ability to work hard is of interest to human biologists and biological anthropologists, if it can be demonstrated that this variation translates into variation in food acquisition, since this is one factor influencing reproductive success.

At the group or population level, it is important to know how work is divided and how resources arising from that work are allocated. In particular, the following factors can influence the distribution of work and resources within a community: (1) group sizes involved in different aspects of the subsistence quest; (2) the division of labour, including gender and age-group differences; (3) the number and age of dependants relative to productive members; and (4) the exchange relations between groups and individuals. In most cases, data are collected at the individual level and aggregated in a variety of ways depending on the type and level of analysis of interest.

In any population, group or community, energy expenditure patterns are a function of: (1) the nature of local energy acquisition, including relevant environmental, technological and exchange factors; and (2) the quantity and patterns of energy consumption. Total energy expenditure consists of energy spent in performing any number of functions, including those necessary to ensure the continued well-being of the group. Furthermore, energy expenditure can play a determining role in limiting and channelling group energy acquisition, principally through the organisation of work patterns and work capacity.

The nature and patterns of acquisition are likely to be energetically parsimonious and adaptive. Consumption patterns and resource allocation within the community or households may also have adaptive significance. Three possible patterns of food allocation could be deemed adaptive, each in different circumstances: (1) preferential allocation to the adult male, if he works hardest; (2) preferential allocation to young children, to buffer them from nutritional stress; and (3) equitable allocation, on the basis of some notion of nutritional adequacy of the diet eaten. Energy expenditure can be deemed adaptive if greater efficiency in performing tasks or improved metabolic efficiency can be demonstrated.

Data on energy acquisition and consumption can be collected at either the group or individual level, while energy expenditure and work capacity can only be measured on individuals and the data aggregated to obtain group values. Because of the impracticability of measuring everyone in a community, particularly when a community is large, studies are usually based on samples that are as representative as possible of whatever aspect of the community the researcher is interested in.

A variety of factors can influence total energy expenditure, including basal metabolism, the level of physical exertion and body size and composition. Related to the level of physical exertion is the possibility that at high levels of performance physical work capacity may be a limiting factor for productivity. Within any group, it is not clear whether the lower energetic cost of maintaining a smaller body size is off-set by lower productivity. Furthermore, it is not clear what the lower limits of smaller body size are, and what the consequences of this might be for reproductive success. However, various authors have established relationships between thinness, productivity and functional impairment. The body mass index (BMI) cut-offs proposed by James, Ferro-Luzzi and Waterlow (1988a) and retained by Ferro-Luzzi *et al.* (1992) for screening for chronic energy deficiency in adults have some meaning in terms of functional impairment; men with BMI below 16 have higher mortality than those with values above 18.5 in India (Reddy, 1991), while both Indians and Ethiopians with BMI below 16 spend less time working than those with BMI above 18.5 (Ferro-Luzzi *et al.*, 1992). In Sarawak, BMI has been shown to be sensitive to reported morbidity in men over the age of 40 years, and women aged 18–40 years (Strickland & Ulijaszek, 1993), while in poor men in Calcutta, BMI below 16 has been associated with greater retrospective morbidity caused by respiratory tract infection and tuberculosis (Campbell & Ulijaszek, 1994).

Human ecology and within-population differentiation

There are a number of ecological questions that can be addressed by examining the energetics of sectors of the community, rather than the community as a whole. Some of these are given in Table 2.4, and include: (1) gender relationships and the way in which they affect the subsistence quest and provisioning for children; (2) changing work expectations of women according to physiological state (pregnant, lactating, non-pregnant, non-lactating); (3) child growth, morbidity and mortality, and the way these influence reproductive performance; and (4) within-group or within-gender work or labour relationships, and their energetic implications.

There are several issues that arise from such within-sector analyses. One of them is whether women work harder than men; also, if workloads are heavy and women work particularly hard, are there ways in which this influences reproductive performance? Leading from this, if heavy workloads are seasonal, is this reflected in seasonality of fecundity and birth rates? Furthermore, if there is seasonality of birth rate, does the need to nurse and maintain infants through their earliest months conflict with the

Table 2.4. *Ecological questions pertaining to sectors of a community*

Gender relationships	Adult male–female differences in: subsistence tasks performed, level of physical exertion, total energy expended in work
	Male–female differences in child provisioning
	Gender preference for child at birth leading to differential male–female mortality rates
	Male–female differences in childrens' involvement in the subsistence economy
Within-sex relationships	Organisation of work groups or hunting or gathering parties, and their energetic implications
Changing roles and expectations in pregnancy and lactation	Extent to which work is reduced in either state Time out of subsistence production after delivering baby
Child growth, morbidity and mortality	Sex preference leading to selected mortality Growth faltering, undernutrition infection and mortality, and its implications for reproductive performance

need to be productive in subsistence activity, particularly in regard to a seasonal fertility boom? Thus, it is possible to pursue various lines of ecological reasoning on the basis of sectoral energetic analyses.

The individual and the group

The relationship between the individual and the group to which they belong is fluid, varies according to the function the individual is performing and is not limited in importance to practical issues of measurement. An example is that of individual and group-work performance (Fig. 2.7). A range of biological, behavioural and environmental factors determine individual work performance, while the way in which the sum of individual performances is used in the group to meet subsistence needs is a function of different behavioural and environmental factors, as well as demographic factors including group size, composition and distribution across space. Therefore, any analysis must include some measure of the relationships between factors operating at the level of the individual and those operating at group level, including within-group variation in ability and output, and the ways in which the group deals with such variation.

One example of the way in which individual performance has been used to examine group relationships is in relation to variation in physical work capacity and within-household variation in energy intakes in one region of Papua New Guinea (Ulijaszek, Brown & Lourie, 1993). Populations in the North Fly region of this country traditionally live in conditions where it is

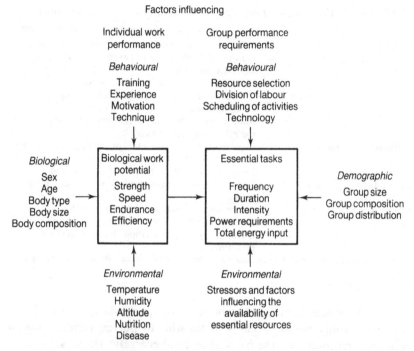

Figure 2.7.   Factors influencing individual work performance and group performance requirements. (From Thomas, 1973b.)

conceivable that physical work capacity ($\dot{V}O_2$max) might be a limiting factor in traditional subsistence practices. As stated earlier, linkages between malnutrition, physical work capacity ($\dot{V}O_2$max) and adult body size have been proposed (Martorell & Arroyave, 1988), and, although it is reasonable to expect that such relationships might influence the food-sharing behaviour and nutritional state of the entire household, this has been difficult to demonstrate. The paper by Ulijaszek *et al.* (1994) is an attempt at such a demonstration.

   Although the Ningerum, Awin and Wopkaimin live at varying altitudes, from 100 to 1600 metres above sea level, and vary in subsistence ecology, their traditional subsistence systems fail to provide adequate intakes of energy and protein (Ulijaszek, 1987) and this is reflected in their smaller body size in comparison with other groups with far greater intakes of energy, such as Wopkaimin males in paid employment (Ulijaszek *et al.*, 1987) and with lower Fly peoples (Ohtsuka, 1983; Hyndman, Ulijaszek & Lourie, 1989).

   Submaximal tests of physical work capacity were carried out on adult

Table 2.5. *Physical work capacity ($\dot{V}O_2$max), body size and energy intakes of adult males, North Fly region, Papua New Guinea*

	Parameters[a] at $\dot{V}O_2$max (ml/kg per min)			One-way ANOVA	
	>40	30–39	<30	F	p[b]
Sample size					
Wopkaimin	3	2	0		
Ningerum	7	10	3		
Awin	5	5	3		
Total	15	17	6		
Age (years)	24.9 (3.9)	27.2 (4.0)	31.0 (4.7)	4.9	*
$\dot{V}O_2$max (ml/kg per min)	45.9 (3.0)	35.5 (3.2)	26.7 (3.6)	89.2	***
Height (cm)	162.4 (4.6)	158.7 (5.0)	158.9 (3.2)	2.8	
Weight (kg)	57.2 (8.9)	53.1 (6.4)	48.3 (4.2)	4.0	*
Body mass index (kg/m²)	21.7 (2.1)	21.1 (2.0)	19.1 (2.4)	3.3	*
Energy intake					
MJ/day	9.71 (2.77)	7.22 (2.62)	6.01 (1.62)	5.9	**
kJ/kg per day	169.9 (34.7)	137.2 (38.5)	124.3 (14.2)	5.3	**

Source: From Ulijaszek *et al.* (1994).
[a] Parameters are given as means with the standard deviation in parentheses.
[b] Probabilities: * $p < 0.05$; ** $p < 0.01$; *** $p < 0.001$.

male volunteers between the ages of 20 and 40 years. In addition, their weights and heights were measured, as were those of all members of the households to which they belonged. Weighed dietary records were collected for all household members above the age of 1 year, and energy intakes estimated. Table 2.5 gives body size and energy intakes of adult males according to higher, middling, or lower $\dot{V}O_2$max category, while Table 2.6 gives these variables for the families of these males, according to the same $\dot{V}O_2$max categories.

Adult males with $\dot{V}O_2$max below 30 ml/kg/min are lighter, with lower BMI and daily energy intakes than those of their counterparts with $\dot{V}O_2$max above this level. Their wives are also shorter and lighter, with lower energy intakes than the wives of men with $\dot{V}O_2$max above 30 ml/kg/min. However, no such differences exist in their pre-adolescent children. Therefore, it appears that these children are buffered against excessively low intakes of energy, at the expense of their parents' energy intakes, regardless of their fathers' physical work capacity. This is supported by the growth and developmental evidence which shows no

Table 2.6. *Body size, energy and protein intakes of adult females and their offspring, according to the male head of household's physical work capacity ($\dot{V}O_2max$)*

A. Wives

	Parameters[a] at husband's $\dot{V}O_2$max (ml/kg per min) of			One-way ANOVA	
	>40	30–39.9	<30	F	p[b]
Sample size					
Wopkaimin	3	2	0		
Ningerum	7	10	3		
Awin	4	4	3		
Total	14	16	6		
Age (years)	22.3 (2.1)	25.1 (4.8)	27.8 (3.7)	4.5	*
Height (cm)	149.9 (5.2)	150.0 (5.7)	140.7 (3.4)	7.9	**
Weight[c] (kg)	42.7 (5.7)	43.2 (3.9)	35.7 (3.9)	6.0	**
Body mass index[c] (kg/m²)	19.0 (2.0)	19.2 (1.1)	18.0 (1.7)	1.3	
Energy intake					
MJ/day	7.23 (1.97)	6.20 (1.78)	4.54 (0.83)	5.0	*
kJ/kg	169.4 (31.8)	141.4 (28.9)	127.2 (20.5)	5.6	**

B. Children aged 1 to 10 years

	Parameters[a] at father's $\dot{V}O_2$max (ml/kg per min) of			One-way ANOVA	
	>40	30–39.9	<30	F	p
Sample size					
Wopkaimin	6	3	0		
Ningerum	9	16	6		
Awin	8	9	5		
Total	23	28	11		
Age (years)	5.8 (2.9)	5.3 (2.8)	6.1 (2.2)	0.4	
Height (cm)	104.3 (15.4)	99.3 (16.3)	99.6 (13.5)	0.7	
Weight (kg)	16.3 (5.0)	14.1 (5.2)	15.2 (4.9)	0.3	
Body mass index (kg/m²)	15.0 (1.0)	14.8 (1.8)	15.3 (1.0)	0.5	
Energy intake					
MJ/day	4.88 (1.66)	4.11 (1.21)	4.11 (0.79)	2.4	
kJ/kg per day	299.6 (36.8)	293.3 (55.6)	270.7 (40.2)	1.4	

Source: From Ulijaszek *et al.* (1994).
[a] Parameters are given as means with the standard deviation in parentheses.
[b] Probabilities: * $p < 0.05$; ** $p < 0.01$.
[c] Non-pregnant women only; sample size 12 for male $\dot{V}O_2$max >40 ml/kg per min and 13 for male $\dot{V}O_2$max 30–39.9 ml/kg per min.

difference in children's heights and weights across their fathers' $\dot{V}O_2$max categories, even though across-category differences are to be found in the fathers and their wives.

This work agrees with an analysis of a combined sample of coastal and highland New Guinea families studied by Ferro-Luzzi *et al.* (1981). They found that the pattern of food distribution favoured children aged 1 to 4 years at the expense of 11 to 17 year olds, pregnant and lactating women and younger women. However, in the North Fly region study, adult males were also nutritionally favoured. Ulijaszek *et al.* (1993) have argued that the body size differences according to $\dot{V}O_2$max categories must start in adolescence, when children of both sexes traditionally become important in the subsistence economy. At this time, the rules that may apply to food sharing with young children no longer do so. The height difference in wives across their husbands' $\dot{V}O_2$max categories could be caused by two factors: (1) food-sharing practices that systematically discriminate against females across the course of adolescence and maturity, as is the case with the Wopkaimin (Hyndman, 1989); and (2) assortative mating, in which the more physically capable males attract brides who are also more physically capable and who may be taller as well as heavier than the average.

Although these linkages need further study, it is possible to extend ecological analysis to consider cross-generational effects, in which households with adult males of average or higher-than-average physical work capacity sustain good or adequate subsistence productivity and energy nutritional status, while households with adult males of very low physical work capacity are in some sense 'energy trapped'; that is, caught in a cycle of disadvantage, where small body size leads to low productivity, low food intakes and poor nutritional status.

Summary

Ecological energetics can be used to examine within-community relationships within an adaptability framework. Although a number of energetics measures can be made at either group or individual level, the two provide different types of information. In particular, the aggregation of data collected at the level of the individual allows variation in different aspects of human energetics, including energy intake and expenditure in bodily maintenance, physical activity and work, reproductive performance and growth, to be examined.

Energy balance is the difference between energy intake and energy expenditure and may vary across season, especially in developing countries. In most countries, there is often long-term energy imbalance, with populations on average either gaining weight across the course of a lifetime, or losing it. There are a number of factors influencing energy expenditure, including body size, physical activity level, body composition, behavioural and cultural factors. Adaptable group energetic responses to environmental change are the product of individual responses and decisions; therefore, it is important to understand energetics at one level to understand the other. An analysis of the linkages between the physical work capacity of adult males and the body size and energy intakes of their wives and families is used to illustrate one way in which individual energetic function has been related to group energetic outcome.

3 Methods

Energetics measures in biological anthropology fall into three broad categories: (1) those that estimate productivity and work output (Ulijaszek & Strickland, 1993b); (2) those that can give some estimate of energy flowing through an ecosystem (Nydon & Thomas, 1989); and (3) those that examine energetic adaptation or adaptability (Ulijaszek, 1995). Of the first category, measures of physical work capacity and energy expended in subsistence activities are important, while of the second, estimates of total energy intake and expenditure and of food energy acquisition are required. Areas of study that fall into the third category include: (1) the energetics of pregnancy and lactation (Ulijaszek, 1993b); (2) the influence of physical activity and energy balance on reproductive function (Ellison, 1991); (3) energetics and human growth and development (Lucas, 1990; Waterlow, 1990a); and (4) metabolic efficiency at different planes of nutrition (James, 1988; Waterlow, 1990b).

Factors that determine the energy expenditure of any group or population include the following: (1) the demographic structure of the group, including age and sex distributions; (2) the patterns of physical activity by different sectors of the population; (3) the proportion of adult women pregnant and/or lactating at any time; (4) the growth and body size of children; (5) the level of maintenance metabolism (equivalent to the BMR); and (6) the energetic efficiency of muscular work.

These factors can influence the energy expenditure of a group in a number of ways. A demographic influence is that of population structure: a group with a large proportion of children below the age of 5 will have lower overall requirements than one with a smaller proportion of young children. In this case, overall group energy requirements will be lower, but the proportion of individuals likely to be engaged in economically productive activity is also likely to be lower. Patterns of physical activity can vary according to the mix of subsistence practices, including paid work, as well as by demographic structure, since some types of work may be better performed in groups rather than by individuals. Also, work may be performed in groups when tasks must be accomplished quickly and efficiently (Panter-Brick, 1993).

The energetic costs of reproductive performance include the additional energy requirements for sustaining a successful pregnancy, and for lactation. This could be large, since in some societies women may spend the majority of their reproductive lifespan either pregnant or lactating. However, this is changing, as total fertility rates in most parts of the developing world are declining (Boserup, 1986). In any population, the level of, and balance between, fertility and infant and child mortality will determine the extent to which the energy costs of reproduction contribute to the overall group energy requirement. Child growth reflects environmental quality, notably nutritional availability, whether it be to the mother or young child (Waterlow, 1988), and experience of, and interaction with, infectious diseases (Tomkins, 1986a, 1988). In developing countries, undernutrition and infection interact, causing linear growth retardation in large proportions of the child population (Martorell, 1985; Martorell & Habicht, 1986; Tomkins & Watson, 1989).

The growth of young infants may be related to the adequacy of dietary energy from breast-milk production. Infants in developing countries who are breastfed to the total exclusion of other foods have a weight velocity in the first month of life which is often in excess of that shown by children in developed countries (Offringa & Boersma, 1987), while lactational performance does not seem to suffer under conditions of moderate dietary restriction (Roberts *et al.*, 1982). Furthermore, the growth of children in general may be related to group energy acquisition, since growth patterns often reflect the overall resource availability to groups or communities. Adult body size and proportion are outcomes of physical growth and development in childhood and can influence productivity and work output, at least when hard physical work is required in subsistence activities (Spurr, 1984).

Aside from the number of adults and children in a population, the physical size of individuals will, to a large extent, determine their energy expenditure in maintenance and physical activity. Furthermore, it is possible that there may be mechanisms whereby maintenance energy metabolism can be lowered, and the energetic efficiency of muscular contraction increased. In such circumstances, there may be important reductions in overall group energy expenditure.

If the aim is to determine the energy relationships between, for example, work input and food energy acquired, then there is no need to go beyond simple energy accounting. That is, simply to measure the number of megajoules expended in work and the number acquired as food. However, more complex models, in which the possibility of energetic stress is acknowledged, require some yardstick with which to determine where such stress might be operating. Yardsticks might include measures of nutritional status that are principally affected by energetic stress, such as: (1) work output as physical activity index (Waterlow, 1986) or physical activity level (James *et al.*, 1988b); (2) estimates of growth faltering against an external reference such as the NCHS growth references (National Center for Health Statistics, 1977); (3) weight loss or negative energy balance in adults (Ferro-Luzzi & Branca, 1993); (4) low weight for a given height in adults, especially low BMI (Ferro-Luzzi *et al.*, 1992); (5) low birthweight (Falkner, Holzgreve & Schloo, 1994); (6) low pregnancy weight gain (Kramer, 1987a,b; Rosso, 1985); (7) poor lactational performance (Abrams, 1991); and (8) low physical work capacity (Spurr, 1988a,b).

Methods of measuring energy intake, expenditure and balance that can be carried out in the field among rural third-world groups and populations are described in this chapter.

Methodological improvements

Recent advances in energetics methodology are considerable and include: (1) the development of less invasive methods for the measurement of total energy expenditure in the field (Haggarty & McGaw, 1988; Prentice, 1990a; Schoeller & Fjeld, 1991); (2) improvements in, and re-evaluation of, older techniques for energy expenditure measurement (Collins & Spurr, 1990); (3) improved understanding of the physiological basis of adaptation to different planes of energy nutrition (Garrow, 1985; James, 1988; James, Haggerty & McGaw, 1988b; Waterlow, 1990a); (4) improvements in energy intake methods (Ferro-Luzzi, 1982; Nelson *et al.*, 1989; Gibson, 1990; Willett, 1990); (5) critiques of the relationship between energy intake and expenditure (Durnin & Ferro-Luzzi, 1982; Durnin, 1990b; Ulijaszek, 1990a); and (6) critical evaluation of field methods for the measurement of body energy stores (Lohman, Roche & Martorell, 1988; Shephard, 1991; Ulijaszek & Mascie-Taylor, 1994; Norgan, 1995a). Furthermore, the energy costs of pregnancy and lactation can be defined with increasing precision (Singh *et al.*, 1989; Goldberg *et al.*, 1993; Guillermo-Tuazon *et al.*, 1992). A review of field energy-balance methods is given by Ulijaszek (1992a).

By partitioning energy expenditure of adults between maintenance and activity and measuring body composition, possible adaptations to low-energy availability can be examined (Waterlow, 1990b), notably in the level (Keys *et al.*, 1950) and efficiency of physical activity (Maloiy *et al.*, 1986; Jones *et al.*, 1987; Jones, 1989; Strickland & Ulijaszek, 1990) as well as down-regulation of basal metabolism (Shetty, 1984; Waterlow, 1986). It is also possible to estimate total energy expenditure (Prentice *et al.*, 1990;

Table 3.1. *Measurement of energy acquisition*

Collection of data
  Regional survey of production
  Questionnaires and conversations concerning production techniques
  Direct measurement of
    production per land unit or animal, per time unit
    food stores
    wastage and loss preceding consumption
    experimentation with productive techniques
  Conversion to caloric values
  Evaluation of the measurement period
    Representativeness of annual production
    Annual variability in production

Source: From Nydon & Thomas (1989).

Spurr & Reina, 1990) and energy expended in physical activity (Durnin, 1990a) by infants and children, and the basal metabolism of infants (Butte, 1990).

In children, the partitioning of energy between growth and physical activity under conditions of low intake has been little studied, although some work has been done in the Gambia (Vasquez-Velasquez, 1988). While examined in the past (Banerjee, Khew & Saha, 1971; Blackburn & Calloway, 1976a,b; Schutz, Lechtig & Bradfield, 1980; Banerjee & Saha, 1981; Nagy & King, 1983), adaptation to varying planes of energy nutritional status have only recently been systematically investigated for pregnant (Durnin, 1988; Forsum, Sadurskis & Wager, 1988; Prentice *et al.*, 1989; van Raaij *et al.*, 1989; Heini, Schutz & Jequier, 1992; Goldberg *et al.*, 1993; Poppitt *et al.*, 1993) and lactating (Sadurskis *et al.*, 1988; Frigerio *et al.*, 1992; Guillermo-Tuazon *et al.*, 1992) women. In addition, a longitudinal study of adaptability of energy expenditure across the course of pregnancy has been carried out with Dutch women (van Raaij *et al.*, 1990).

Measuring energy acquisition

This can be done at a number of different levels, according to the scale of the study in hand. Tables 3.1 and 3.2 summarise the various procedures which can be used. Regardless of the method used, all require the conversion of quantitative information into energy values. It is important to understand the representativeness of values collected, in terms of annual variability in production.

Data can be collected by: (1) utilising data collected at regional level by government officers employed in agriculture departments; (2) using questionnaires, either self-administered by the subjects of the study, or by an interviewer; or (3) direct measurement. Of these, direct measurement is often the most reliable, though most time-consuming for the researcher and prone to uncertainty; that is, by simply being present, the observer may influence what is being measured. Regional data often do not exist for subsistence agriculture, although they may for cash-cropping; for pastoralist or hunter–gatherer populations, government officers concerned with agricultural development may simply ignore their existence, and data will be lacking as a consequence. Questionnaire methods may work where people have some measure of yields. Yield estimates need not be in weight, but it must be possible to convert values into metric units. For example, yields given in bags of rice, or stands of maize, can be converted into kilograms or tonnes by weighing representative specimens of the measure in question, taking care to note any variability in the unit of measurement. However, not all subsistence cultivators may be aware of their yields, although upland rice cultivators in Sarawak (Strickland, 1986), field rice cultivators in Nepal (Strickland *et al.*, 1993) and sago cultivators in Papua New Guinea (Ulijaszek & Poraituk, 1983) can make accurate statements about yields of their main staple crops.

There are a number of problems associated with the conversion of crop or food weights into energy values, the most important of which is variability in the composition of foods. In preparing food composition tables, compilers attempt to produce values that are representative of the food item in general, and, therefore, the tables cannot be expected to predict accurately the composition of a single sample of any given food (Southgate, 1993). Although errors caused by this factor might balance out if the diet is composed of a large number of food items, this is unlikely to happen if one staple crop forms the vast majority of the diet, or the complete harvest of any dominant crop is being assessed. There is also variability in the energy content of different types of the same food crop. As an example, Table 3.3 gives the energy, protein and water content for a staple food used widely in Papua New Guinea, the sweet potato (Norgan, Durnin & Ferro-Luzzi, 1979). There are two points to note: (1) energy content varies according to the method of preparation; and (2) the water content can affect the calculated energy value. Therefore, it is important to record the state of the food crop as accurately as possible; was it cooked, or raw? If cooked, how was it prepared? Furthermore, if there is a dominant staple crop and the facilities exist in the proximity of the study, it may be worthwhile taking samples for water-content estimation. This is because correction for differences in water content between the study samples and

Table 3.2. *Uses and limitations of commonly used methods to assess energy intakes*

Method	Procedure	Uses	Limitations
Duplicate portion analysis	All food eaten over a defined period is examined, weighed and duplicated; the duplicate sample is then analysed for energy value by bomb calorimetry and other methods	Accurate energy balance studies under controlled conditions	Extremely time consuming and expensive of laboratory resources. Because a duplicate portion is required, energy intake is unlikely to represent habitual intake
Weighed food intake	All food eaten over a defined period is weighed. Samples of some commonly occurring foods may be taken for energy content analysis for use in subsequent analyses. Otherwise, energy intakes are calculated from weights of foods eaten and literature values for the energy content of foods	Assessment of actual, or usual intakes, depending on number of days of measurement and level of seasonality of food intake	Time consuming and expensive of staff. Subjects may change their usual eating patterns to simplify weighing or to impress the investigator with how poor, or how good, their eating habits are. Often difficult to determine the extent and direction of the bias, unless intakes are so low that they cannot sustain energy balance
Estimated food record	Record of all food and beverages consumed over 3 to 7 days is taken. Quantities estimated from household measures; energy intakes estimated from literature values for the energy content of the foods eaten	Assessment of actual, or usual intakes, depending on number of days of measurement and level of seasonality of food intake	Accuracy depends on conscientiousness of subjects and their ability to estimate quantities. Longer time frames result in a higher respondent burden and lower cooperation
24-hour recall	Subjects' food intake over the previous 24 hours is recalled in an interview. Quantities are estimated in household measures using food models as memory aids and/or to assist in quantifying portion sizes. Energy intakes calculated from estimated weights of foods eaten and literature values for their energy content	Assessment of average usual intakes of a large group, provided that the sample is representative and that the days of the week are adequately represented	Low-respondent burden, but single 24-hour recalls are unlikely to give estimates of usual intakes for individuals. Aggregated group average cannot be used to examine variation of energy within a population

Source: From Acheson *et al.* (1980); Ferro-Luzzi (1982); Gibson (1990).

Table 3.3. *Reduced variation in the reported energy content of sweet potato after correcting for differences in water content between different cultivars and modes of preparation*

Sample number	Variety	Method of preparation	Content[a] without correction for water content		Content[a] corrected for water content	
			Energy (%)	Protein (%)	Energy (%)	Protein (%)
1	1	Mumu'd	112	127	103	116
2	1	Roasted	136	182	102	137
3	2	Boiled	109	236	102	223
4	3	Boiled	89	73	98	80
5	3	Mumu'd	106	173	101	164
6	3	Roasted	108	118	102	112
7	4	Boiled	93	55	102	59
8	4	Mumu'd	113	145	101	131
9	4	Roasted	93	136	101	148
10	5	Roasted	65	54	95	80
11	6	Boiled	85	82	102	98
12	6	Mumu'd	77	137	101	178
13	6	Roasted	126	164	105	136
14	7	Mumu'd	93	183	97	190
15	8	Boiled	89	91	97	99
16	8	Mumu'd	98	117	99	120
17	8	Roasted	103	109	103	109
18	9	Mumu'd	91	73	101	81
Mean			101.8	125.3	100.7	125.6
Standard deviation			23.9	48.4	2.4	41.7
Coefficient of variation			23.5	38.6	2.4	33.2

Source: From Norgan *et al.* (1979).
Mumu'd: cooked in a traditional New Guinea 'pit-oven'.
[a] Energy and protein contents of 19 samples of sweet potato are expressed as a proportion of a further sample whose water, energy and protein contents are 68.2 g, 493 kJ and 1.1 g/100 g, respectively.

the food composition table values can greatly reduce the error of estimation.

Care should be taken over the use of food composition tables. Since the first compilation of such tables (Konig, 1878), a large number of compilations have become available for use in a great number of countries and regions. Extensive lists of these can be obtained by consulting either the International Directory of Food Composition Tables (Infoods, 1986), or Arab, Whittler and Schettler's (1987) *European Food Composition Tables in Translation.* However, very few of them give any measure of variability around the reported mean values. Such data would be valuable in determining errors in estimates of energy intake or production which might be caused by variability in the energy content of the reference foods given in the food composition tables. The energy values of foods are the sum of their carbohydrate, fat and protein contents (in the absence of alcohol). Food composition tables predict total available carbohydrate and protein intakes with reasonable accuracy, while fat is poorly predicted, at least in western countries (Southgate, 1993). The greater the fat content of the diet, and the more variable the sources of fat, the less accurate is the prediction of total energy intake from the records obtained of foods used or obtained.

Food composition tables rarely manage to give a complete coverage of foods likely to be used by a group or population. This is because the majority of the intake of most nutrients is provided by relatively few foods, and compilers of food composition tables focus their resources on those foods (Greenfield & Southgate, 1992). It is important to assign values to all foods reported as consumed or produced, since omission will lead to under-estimation. When published values are not available in the food composition table of choice, energy content can be assigned by either: (1) using values for similar foods given in that table; (2) using values for that food, from a different food composition table; or (3) giving them the average composition of the remainder of the diet. Option (3) is the least preferred. Another source of error in the use of food composition tables lies in the reporting of energy values in the tables themselves. Values reported in tables are for metabolisable energy and are derived from the proximate composition of the food using energy conversion factors (Southgate, 1993). These factors vary from table to table, introducing the possibility of error if more than one set of tables is used in a particular analysis.

When assessing food production, there may be problems in determining the representativeness and variability of annual production. In agricultural communities, measuring the harvest of each crop will be representative of the entire year if harvesting is clearly delineated by a season, such as with rice, maize, wheat, sorghum or millet cultivation. Note should be made of whether or not there is a second, smaller harvest at another time of year for any of the crops observed. With crops that do not have such a clearly delineated harvest, or may be harvested across the year, the issue of representativeness of the study period is germane; indeed, the question of whether or not the harvesting of a particular crop is similar all year round can be difficult to answer with staples such as cassava, sweet potato or plantain, none of which can be stored post-harvest for prolonged periods. In pastoralist communities, there is seasonality of milk production; Fig. 3.1 shows the variation in milk output in different animals used by the Tuareg pastoralists in Mali (Swift, 1981). There may also be seasonal variation in

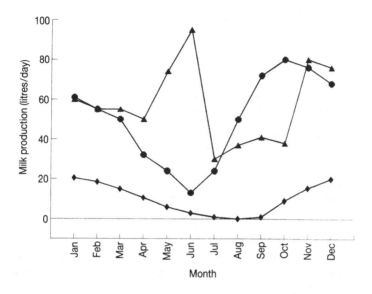

Figure 3.1. Milk production in a flock of 25 goats (♦) and in herds of 25 camels (●) and 25 cattle (▲), Tuareg pastoralists, north Mali. (From Swift, 1981.)

the energy content of the milk, since both quantity and quality of milk output are a function of the quality of the forage available, and of the quantity of water available.

Energy intake

There are a number of techniques available for measuring energy intake (Tables 3.2 and 3.4). Regardless of the method used, data must be able to give estimates of weights of different foods eaten, either in households or by individuals. These weights must then be converted into energy values using food composition tables, acknowledging the problems attendant with their use. More detailed examination, beyond the level of the group or the household, requires energy intake data to be disaggregated by sex and age group. Furthermore, intakes may be compared with international dietary references, for the presence or absence of notable dietary energy deficiency, either from observations of clinical pathologies, or from anthropometric measures such as weight for height or age, and skinfold thicknesses (Jelliffe & Jelliffe, 1989; Gibson, 1990). Although the corroboration of food intake with other lines of evidence for nutritional stress might seem appropriate, there are considerable problems with this approach: (1) estimates of energy intake are associated with much greater error than measures of energy

Table 3.4. *Procedures for estimation of energy consumption*

Collection of data
  Weighing of food consumed
  Questionnaires concerning food use
  Daily recording of food items consumed
Conversion to caloric values
Establishment of consumption by sex or age group
Evaluation of the adequacy of energy consumption
  Comparison with international standards
  Presence of deficiency-related symptoms
  Anthropometric indicators of caloric balance

Source: From Nydon & Thomas (1989).

expenditure and may be prone to underestimation (Ulijaszek, 1992a); and (2) there is no ideal reference for nutritional assessment (Ulijaszek & Strickland, 1993b).

Collection of energy intake data

There is no single, ideal method for estimating energy intake (Gibson, 1990), and a huge literature has been generated in attempts by numerous authors to address problems associated with dietary intake measures (Balogh *et al.*, 1968; Morgan *et al.*, 1978; Ferro-Luzzi, 1982; Jain *et al.*, 1982; Durnin & Ferro-Luzzi, 1982, Stuff *et al.*, 1983; Yarnell *et al.*, 1983; Willett *et al.*, 1985; Pietinen *et al.*, 1988; Boutron *et al.*, 1989; Nelson *et al.*, 1989; Durnin, 1990b; Bingham, 1991; Cole, 1991; Huss-Ashmore, 1995). Volumes on nutritional epidemiology written by Willett (1990) and edited by Margetts and Nelson (1991) provide detailed accounts of these issues in the epidemiological context. The rest of this section will give descriptions of intake methodology appropriate to anthropological research.

There is no merit in using a more elaborate or expensive method than is needed for the purposes of the study; the more detailed the desired data, the more time consuming and expensive is the method needed (Ferro-Luzzi, 1982). No dietary method is free from error, the more detailed methods being subject to types of error that are different from the less detailed methods. Table 3.2 gives the uses and limitations of some of the most appropriate methods for the assessment of energy intake.

Household measures of food consumption are among the most accurate, since they are among the least invasive of methods. However, household-level data does not allow disaggregation by age and sex. Measures of individual intakes are needed to do this; of the methods available, weighed food intake is potentially the best, although the potential for the food consumption behaviours of the subjects to be altered during the period of study is great. Certainly, the more invasive the method, the more likely is behaviour to be modified (Ferro-Luzzi, 1982). However, less-invasive methods such as 24 hour recall are likely to result in under-reporting, at least in less-industrialised countries. Comparisons of the 24 hour recall method with weighed food intake have shown the latter to give lower estimates of energy intake in Jamaicans (Fox, Campbell & Lovell, 1968) and rural Malawian (Ferguson *et al.*, 1989) and Ghanaian children (Ferguson, Gibson & Opare-Obisaw, 1994).

In any study, a balance must be sought between the desire to maximise accuracy and to keep the duration of study to a minimum. Operating within these constraints, researchers should use equations available for the estimation of the reliability of energy intake (Ferro-Luzzi, 1982; Cole, 1991) to arrive at their own estimates of reliability, having decided on the number of days of observation per subject and the sample size possible, after collecting the data. Such measures of reliability are of great value in the analysis and interpretation of human energetics data.

In addition to problems associated with the use of dietary intake methods, there are other considerations, such as the sampling method and the need to disaggregate the data by age and sex, which can influence estimates of energy intake. Furthermore, food preparation techniques can affect the energy content of foods eaten, while the under-recording of substances which the subject may not regard as food but which contain energy can result in errors of estimation.

The choice of the sampling period is an important consideration in carrying out dietary studies. Human groups exhibit a number of different food consumption patterns; very few of them show no variation in quantity or types of food eaten either from day to day, or across seasons. Worldwide, patterns of food consumption are almost universally related to the patterns of subsistence. For example, in Papua New Guinea, Ningerum horticulturalists show no variation in intake, but enormous differences in the sources of dietary energy across seasons (Ulijaszek, 1985). In Senegal, Serere pastoralists show seasonal variation in energy intake and in types of food used (Rosetta, 1988a). This is related to the type of agro-pastoralism that these people practise. Mixed horticulturalists in Chad and Cameroon (de Garine and Koppert, 1990), and agriculturalists in the Gambia (Fox, 1953) engage in post-harvest gorging to a greater and lesser extent, respectively.

In estimating the energy intake of a group of people, it is important to use a sampling period which is representative of longer-term habitual intake, or of intake in a particular season. In developing countries, seasonal variation in food availability may suggest that sampling be carried out at different times of year, representative of the different seasons. Extrapolating from data which have not taken into account seasonal bias can lead to erronous generalisations about the group in question (Ulijaszek, 1992b).

Without carrying out prior investigations, it is often difficult to determine the most appropriate sampling period and protocol. There are various ways of arriving at some approximation, however. Ethnographic, social and nutritional literature may give some clues as to variation in dietary patterns; alternatively, researchers and government officials who have worked with the population of interest may be able to provide important information. Rainfall and ambient temperature figures can give an indication of the extent of climatic sesonality experienced by rural subsistence-oriented populations, which may translate into seasonality of food intake and, therefore, of energy.

In some populations, the consumption of non-food items could be important. In particular, alcoholic beverages contain energy; subjects consuming even only small amounts of alcoholic drink that are not reported may have their energy intakes significantly under-reported. One cannot make generalisations about the contribution of alcohol to the diet, even of apparently similar groups; for example, the consumption of alcohol by Finnish and American farmers represents about 1% of their total daily energy intake, while for Italian farmers it forms about 15% of daily intake (Ferro-Luzzi, 1982).

The weighed dietary record has in the past been regarded as the reference method of dietary methodology (Bingham, 1991), but it is only recently that the validity of the method for assessing the habitual diet of free-living individuals has been questioned (Prentice *et al.*, 1986b; Livingstone *et al.*, 1990; Ulijaszek, 1992a; Huss-Ashmore, 1995). Comparisons of mean energy intakes with expenditures in a number of studies in non-industrialised nations show that in six out of twelve samples, reported intakes are more than 1 MJ below expenditure, a difference which cannot be explained on the basis of possible seasonal fluctuations in work output and food availability (Fig. 3.2). A comparison of energy intake and expenditure in studies where expenditure was measured by the doubly labelled water method confirms this view (Huss-Ashmore, 1995). Furthermore, 24-hour urinary nitrogen, first proposed by Isaksson (1980) as an independent validity check on dietary survey methods, has been used in Cambridge, England to show that 39% of subjects involved in a 16-day weighed dietary intake study significantly under-reported protein intake and, therefore, dietary intake in general. Simultaneous use of both the 24-hour urinary

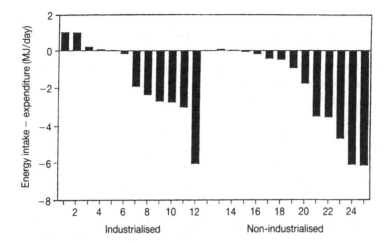

Figure 3.2. Difference between mean energy intake and expenditure from a number of studies where both measurements were made on the same subjects. 1. Male medical students, Singapore, $n=7$ (Saha, Tan & Banerjee, 1985); 2. Female medical students, Singapore, $n=16$ (Saha *et al.*, 1985); 3. Elderly, Arizona, $n=38$ (Vaughan *et al.*, 1991); 4. British soldiers, $n=12$ (Edholm *et al.*, 1955); 5. Younger subjects, Arizona, $n=64$ (Vaughan, *et al.*, 1991); 6. Females, California, $n=5$ (Mulligan & Butterfield, 1990); 7. Females, Northern Ireland, $n=15$ (Livingstone *et al.*, 1990); 8. Non-obese adolescents, Boston, $n=28$ (Bandini *et al.*, 1990); 9. Female runners (54 km/week), California, $n=7$ (Mulligan & Butterfield, 1990); 10. Laboratory personnel, Chicago, $n=4$ (Schoeller & van Santen, 1982); 11. Males, Northern Ireland, $n=16$ (Livingstone *et al.*, 1990); 12. Obese adolescents, Boston, $n=27$ (Bandini *et al.*, 1990); 13. Male farmers, Burkina Faso, $n=11$ (Bleiberg *et al.*, 1981); 14. Women, Swaziland, $n=16$ (Huss-Ashmore *et al.*, 1989); 15. Non-lactating women, Guatamala, $n=5$ (Schutz *et al.*, 1980); 16. Women in the planting and cultivating season, Guatamala, $n=11$ (McGuire & Torun, 1984); 17. Males on the British Antarctic Survey Base, Antarctica, $n=12$ (Acheson *et al.*, 1980); 18. Lactating women, Guatamala, $n=18$ (Schutz *et al.*, 1980); 19. Women in the non-agricultural season, Guatamala, $n=11$ (McGuire & Torun, 1984); 20. Female farmers, Burkina Faso, $n=14$ (Bleiberg *et al.*, 1981); 21. Male farmers, the Philippines, $n=9$ (de Guzman, 1984); 22. Housewives, the Philippines, $n=10$ (de Guzman, 1984); 23. Female shoemakers, the Philippines, $n=10$ (de Guzman, 1984); 24. Non-pregnant non-lactating women, the Gambia, $n=4$ (Singh *et al.*, 1989); 25. Lactating women, the Gambia, $n=9$ (Singh *et al.*, 1989).

nitrogen and doubly labelled water methods has shown that both identify the same subjects who under-report their food intake when asked to keep weighed records (Black, Jebb & Bingham, 1991). This suggests that the urinary nitrogen method of validation may be appropriate for anthropological work, since it is quite cheap, relatively simple to do and is a long-standing biochemical method of assessment of protein nutritional status (Gibson, 1990). It remains to be seen, however, whether the same precision of identification is obtained in populations with habitually low protein intakes. Moreover, the method assumes that subjects are in nitrogen balance, with no loss or gain in bodily protein. This would suggest that the method is unlikely to be accurate in groups that habitually experience seasonal changes in body weight, caused by variation in food intake.

Another way of checking for abnormally low, non-habitual levels of reported energy intake is to compare reported intake with a multiple of BMR, either measured or predicted from body weight using regression equations, such as those of Schofield (1985) for European populations, or those of Henry and Rees (1991) for tropical populations. Energy intakes of less than $1.27 \times$ BMR cannot be true, except under conditions of extreme negative energy balance, in which case this level is not habitual. Habitual values below this level are incompatible with life (FAO/WHO/UNU, 1985) and, in most cases of dietary intake, are likely to result from gross under-reporting or under-eating by the subject, or both.

Energy expenditure

Procedures for estimating energy expenditure are summarised in Table 3.5. In any study, some but not all of these measures will be employed; furthermore, the methods chosen will depend on the questions in hand and the level of precision and portability of the equipment needed. These issues are dealt with in some detail in reviews of human energetics methods by Ulijaszek (1992a), Norgan (1995b) and Leonard (1995). The measurement of energy expenditure under free-living conditions may involve the use of any of several techniques, including oxygen uptake, activity diaries, heart-rate monitoring and isotopically labelled water.

It may be important to determine which of all the activities performed are the most important in the subsistence pattern, the performance effort necessary to carry them out and the endurance capacity of individuals in performing subsistence activities. The energetics of work performance, such as measures of physical work capacity, are important in this regard. Furthermore, it is usually important to know how subsistence tasks and labour are divided between the sexes, and the extent to which children are involved in economic activity.

The importance of using more than one measure

Methods for assessing energy expenditure under field conditions are summarised in Table 3.6. Although the use of a single method often provides valuable information, it is more profitable to use two or more

Table 3.5. *Procedures for estimation of energy expenditure*

Survey of habitual activity
Determination of critical activities
   Importance in subsistence pattern
   Extent to which activity is relied on
   Performance effort necessary
Determination of participants in activities
Measuring energy expenditure
   In field measurements
   Under standardised testing conditions
Estimation of the energy cost of activities not examined
Time–motion studies
Assessment of endurance capacity in performing activities

Source: From Nydon & Thomas (1989).

methods in association with each other. For example, the use of doubly labelled water can give TEE values averaged over a period of days. Additional measurement of BMR allows this expenditure to be partitioned into maintenance expenditure, and total energy expended in physical activity plus TEF. Often, the TEF is ignored when attempting to index levels of physical activity; rather, it is generally assumed that TEF does not vary significantly between individuals. Despite the inaccuracy of this assumption (Horton, 1983), the TEE of an individual divided by their BMR is a better measure of PAL in adults than most (James *et al.*, 1988a), since it is standardised for both body mass and biological variation in metabolic rate (Waterlow, 1986). In order to examine day-to-day variation in activity level, the use of activity diaries or heart-rate monitoring in association with classical calorimetric techniques are appropriate for estimating TEE, since isotopic methods can only give values for TEE averaged across a number of days.

A growing number of publications review the principles underlying techniques of calorimetry (Blaxter, 1989; McLean & Tobin, 1987; Norgan, 1995b), diary methods and heart rate monitoring (Collins & Spurr, 1990; Ulijaszek, 1992a; Leonard, 1995; Norgan, 1995a,b) and isotopic labelling (Coward, 1988; James *et al.*, 1988b; Prentice, 1990b; Ulijaszek, 1992a,b). Short descriptions of field methods appropriate for anthropological and human biological study are now given.

Activity diaries

These may be self-recorded or may be recorded by an observer who is allocated to the subject. Although it is desirable to record activities for

Table 3.6. *Uses and limitations of methods used in the measurement of energy expenditure*

Method	Procedure	Uses	Limitations
Indirect calorimetry	Rate of oxygen consumption estimated from gas analysis of expired air and the total volume of gas expired over set period of time	Measurement of basal metabolic rate, energy cost of different activities	Size and portability of equipment may influence the range of activities that can be measured. Possible inaccuracies caused by leaks and gas diffusion from sampling bags; inaccuracies in flow meters. Use of mask or respiratory valve hampers some activities and makes participants self-conscious
Time allocation	Recording of time spent performing activities in the course of the day either continuously, or on a minute-by-minute, or 5 minute-by-5 minute basis	Overview of different activities performed, some of which may be followed up with more precise techniques. In association with calorimetry, it can give a measure of daily TEE and energy expended in key activities. In association with calorimetry and body composition measures: energy cost of reproduction, growth can be estimated	Observation and recording may interfere with habitual activity
Heart rate monitoring	Minute-by-minute transmission, recording and storage of heart rate by electronic device fitted to subject at the start of the day. Down-loading of heart rates onto computer prior to analysis	In association with calorimetry and possibly also activity diaries: it can estimate daily TEE, energy cost of daily physical activity and energy cost of key activities. In association with calorimetry and body composition measures: energy cost of key activities, e.g. reproduction, can be measured	Inaccurate at low or extremely high levels of physical activity

| Doubly labelled water | Dose of known volume of stable isotope administered to subject by mouth. Urine or saliva samples taken every day over a 7–10 day period, or once at the beginning of the study and again at the end | Daily TEE. In association with calorimetry: TEF, energy cost of reproduction, growth can be measured | Does not give a measure of day-by-day variation in total energy expenditure; costly |

Source: From McLean & Tobin (1987); Collins & Spurr (1990); Prentice (1990b).

every minute of the waking day, this is usually impossible. Such a recording schedule can stretch the patience of the recorder, while also being impossible for the subject to self-record, since the practice of recording will interfere with their habitual activity. Observations made every 5 minutes are usually more practicable but result in a loss of information. However, this may not matter if the diaries are to be used for energy expenditure estimates only and not for the study of behavioural ecology requiring extremely fine-grained data. A diary should be formulated in which daily activities are recorded using a simple code; activities should reflect the range of activities normally undertaken by subjects and not be taken piecemeal from a methods manual. Where possible, a short pilot study should be carried out to ensure that the range of activities chosen is appropriate and that the study is feasible.

Diary methods are more feasible in rural communities than in urban ones. In urban settings, subjects may interact with a larger number of individuals on a daily basis than subjects in rural communities and may be reluctant to have an observer with them everywhere they go, since this may be a source of embarrassment. In both settings, however, the presence of the observer may affect the observation, regardless of whether they are a local person or not. Subjects may decide to limit their activities on the days of observation by either staying at home or going to visit relatives or friends. It is, therefore, important to build up some picture of habitual activities in the community prior to the study against which to compare the activity diaries of the individuals studied (Ulijaszek, 1992a).

Records need to be taken for 3 to 7 days, since a day-to-day variation in mean activity level of $\pm 30\%$ is not uncommon (Collins & Spurr, 1990). Where subjects are engaged in paid employment, the period of observation should include days off as well as working days. Direct measurements of energy expenditure of the individual for each or most of the activities recorded, or obtained from appropriate tables of energy costs of activities (Edholm *et al.*, 1955; Passmore & Durnin, 1955; Durnin & Passmore, 1967; Brotherhood, 1973; Norgan *et al.*, 1974; Bleiberg, Brun & Goihman, 1980; Brun, Bleiberg & Goihman, 1981; Brun *et al.*, 1988; Lawrence & Whitehead, 1988; Inoaka, 1990; James & Schofield, 1990), can then be used to estimate total energy expenditure for the period of observation. A list of the energy costs of a range of activities is given in Chapter 2 (Table 2.3, p. 35).

The energy cost of any given physical activity can vary enormously, either between subjects in the same study, or between mean values published by different authors (Brun, 1992). Table 3.7 gives the range of energy cost of activities performed by female farmers in tropical China

Table 3.7. *Energy cost of physical activities performed by female farmers in tropical China*

Activity	n	Mean (kJ/min)	Range (kJ/min)	Maximum/minimum ratio
Lying inactive	3	3.4	3.0–3.7	1.24
Sitting inactive	11	4.5	2.8–6.7	2.41
Standing resting	4	6.0	4.5–7.0	1.55
Squatting, washing clothes	4	8.7	7.8–10.2	1.31
Standing, sowing rice	10	9.0	7.8–11.4	1.46
Bending, harvesting potatoes	8	9.9	8.2–11.9	1.45
Squatting, bundling rice	6	10.1	8.4–12.9	1.54
Walking, tapping rubber	5	10.5	7.4–12.8	1.74
Bending, transplanting rice	31	11.9	7.5–17.9	2.38
Ploughing with buffalo	4	12.3	10.4–15.0	1.44
Bending, cutting rice	26	13.5	5.2–17.9	3.45
Bending, planting potatoes	7	14.2	10.5–16.9	1.61
Walking, carrying 30–35 kg	5	15.7	12.8–19.5	1.52
Standing, hoeing	3	16.0	12.9–21.3	1.65
Standing, threshing rice	8	16.6	10.0–22.3	2.23

Source: From Brun *et al.* (1988).

Table 3.8. *Mean energy cost of comparable physical activities performed by female farmers in China and the Gambia*

Activity	Energy cost (kJ/kg per 24 hours)	
	Tropical China	The Gambia
Sitting inactive	138	150
Standing resting	183	151
Squatting, washing clothes	268	394
Bending, harvesting potatoes	303	229
Bending, cutting rice	413	248
Standing, hoeing	490	451

Source: From Brun *et al.* (1988); Lawrence *et al.* (1988b).

(Brun *et al.*, 1988), while Table 3.8 compares mean values from this study with those for comparable activities for women in the Gambia (Lawrence *et al.*, 1988b), standardised for differences in body weight.

Although the ranges reported in Table 3.7 may in some cases be a function of small sample size, the range between lowest and highest energy expenditure for any given task is extremely variable. In this example, differences between the highest and lowest values for the activities with the lowest and highest variation are 24% and 345%, respectively. A large number of reference values for the energy cost of subsistence activities are based on very small sample sizes, and caution is advised in the interpretation of energy-expenditure values derived from this method. Furthermore, the range of variation seems to be independent of the mean level of energy expenditure associated with any particular task: that is, an activity requiring harder work in general is not performed at different levels of effort by different individuals, at least in tropical Chinese farmers. This is also true of women engaged in sago making in Papua New Guinea (Ulijaszek & Poraituk, 1993) and is likely to hold true for activities which are organised into work parties, especially where activities are synchronised.

Table 3.8 shows that even when energy expended in physical activities is standardised for differences in body size, there are differences in mean values for the same task in two different groups of women. Again, the sizes of the differences do not appear to be related to the absolute level of effort required to perform the task, nor are the differences all in the same direction; that is, Chinese women expend more energy in doing some tasks, Gambian women in doing others. It is unlikely that the majority of these differences could be attributed to differences in the technology used to perform a given task, although this might be a contributing factor in hoeing, harvesting potatoes and cutting rice. Differences in technique and posture are likely to be far more important and should be considered when using this approach to energy-expenditure measurement. Despite its history as a method for estimating energy expenditure, it is clear that more work is required to improve the accuracy of estimation by the activity diary method. However, at its most basic, it can be cheap, requires no more equipment than a timer, a pen and paper and gives acceptable values at group or population level, although observers should be aware that measures of energy expenditure by this method are approximate.

Twenty-four hour energy expenditure values can be obtained from activity diaries by assuming that the remainder of the day was spent sleeping, and that the metabolic rate while sleeping is the same as BMR. Although it is known that the energy expenditure of sleep is about 5% lower than BMR (Garby *et al.*, 1987; Goldberg *et al.*, 1988), this would give an over-estimate of energy expenditure of only about 1% in an adult male weighing 70 kg and sleeping for 8 hours out of 24. In practice, subjects may have energy expenditures above a defined sleeping level for part of the night, if they wake up or are restless. Using BMR in calculating the energy expended during the night gives a 5% allowance for such possibilities. Overall, this level of error is considered by Goldberg *et al.* (1988) to be negligible.

Calorimetric methods

Although there is a wide range of equipment available for estimating energy expenditure from oxygen consumption (McLean & Tobin, 1987), bag systems and portable open-circuit systems of greater or lesser sophistication are the most widely used in the field.

The collection of expired air for the purposes of gas analysis and estimation of human energy expenditure was proposed by the English physiologist Douglas in 1911. The method is a simple one, but one which has stood the test of time. Briefly, the subject breathes into the Douglas bag by way of a mouthpiece while performing a set activity for a known period of time (usually more than 5 minutes). Their rate of oxygen consumption at standard temperature and pressure is then calculated from the difference in oxygen concentration between expired and atmospheric air. Energy expenditure is calculated using some value of heat production per litre of oxygen consumed in the metabolism of the macronutrients carbohydrate, fat and protein. This is the respiratory exchange ratio (RER) and it varies according to the mix of these nutrients being metabolised at any given time (Brody, 1945; McLean & Tobin, 1987).

Someone consuming a diet containing a mix of macronutrients would typically have an RER of around 0.85. If this value is assumed for an individual being measured and they are untypically eating a diet consisting predominantly of only one of the macronutrients, then an error of measurement of up to $\pm 4\%$ is possible. In practice, such deviation is unlikely, and the level of error caused by variation in RER away from the assumed value is considered to be within an acceptable range for field measurements of energy expenditure.

Other major sources of error in the Douglas-bag technique include leaks from the bag and diffusion of gas through the walls of the bag (Perkins, 1954). While one should regularly check the bags in the course of field work, the problem of elevated oxygen measurements as a result of outward diffusion of carbon dioxide is minute, providing that contents are analysed within 30 minutes of collection (Collins & Spurr, 1990).

Although the availability of portable open systems for estimating oxygen consumption has the potential to revolutionise the study of energy expenditure in field conditions, caution is advised. A number of compact devices exist, but each has limitations, and they vary in reliability. The following items are the more useful of the equipment available: (1) the Kofranyi–Michaelis (K–M) meter or Max Planck respirometer (Kofranyi & Michaelis, 1940); (2) the portable respiratory air monitor (Harvard Instruments); and (3) the Oxylog (Humphrey & Wolff, 1977).

Heart-rate monitoring

There is a close linear relationship between heart rate and energy expenditure when they are both above resting levels (Collins & Spurr, 1990). The method has been tested against total body calorimetry (Avons *et al.*, 1988; Spurr *et al.*, 1988; Ceesay *et al.*, 1989) and doubly labelled water (Livingstone *et al.*, 1990) and has been shown to give group mean values for total daily energy expenditure which vary between -0.6 and $+6.8\%$ of reference values.

Early attempts at using heart-rate monitoring for measuring energy expenditure were doomed to failure for several reasons. Heart-rate accumulators recorded heart rates across the day from which an average daily heart rate could be obtained. Since this value is usually only a few beats above resting values, estimates of expenditure were poor, reflecting the poor relationship between these variables at low levels of physical activity. Recorders currently available are capable of storing heart rate values on a minute-by-minute basis for 16 hours or more. These data can then be down-loaded onto computer and subsequently analysed. The low level of invasiveness of the method and the availability of portable lap-top computers for down-loading has made this an attractive method for use even in remote areas, provided that subjects are reasonably active. This method has been used with success in rural Nepal (Ulijaszek, Tuffrey & Strickland, 1994) and among Siberian pastoralists (Leonard, 1994).

Although the relationship between heart rate and oxygen consumption is linear above a certain value, it is different for each individual, and may change with altered levels of fitness (Astrand & Rodahl, 1986). It is, therefore, important that a calibration curve be created for each individual to be studied, at the time of study. If the purpose of the study is to examine seasonality of energy balance or physical activity, such calibration should be obtained at each season of study, if for any reason it is likely that the subjects may show seasonal variation in physical fitness.

The calibration curve can be obtained in the following manner. Low level, resting energy expenditure is calculated from oxygen consumption using a standard method such as the Douglas bag, while lying, sitting and standing quietly. Heart rate is concurrently measured. The subject then exercises at several progressively increasing workloads, while oxygen consumption and heart rates are measured. This provides the regression line for converting heart rates into oxygen consumption and thence energy expenditure. Figure 3.3 gives the calibration curve for a fit young individual.

The critical heart rate (FHFLEX) below which the linear relationship

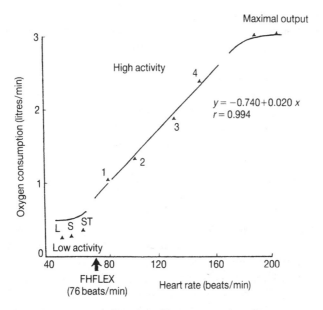

Figure 3.3. Calibration of heart rate against oxygen consumption for one subject. L, lying; S, sitting; ST, standing; 1–4, increasing levels of exercise; FHFLEX, critical heart rate.

breaks down needs to be determined for each individual. A method that has given reliable results has been to take the mean of the highest heart rate during the resting measurements and the lowest during the exercise measurements (Spurr *et al.*, 1988; Ceesay *et al.*, 1989).

Total energy expenditure can be calculated by partitioning activity into three time groups: (1) time spent above FHFLEX; (2) time spent below FHFLEX; and (3) time spent sleeping. Computer analysis of heart-rate data involves a comparison of each minute's heart rate against the subject's FHFLEX. When above FHFLEX, oxygen consumption is calculated from the heart-rate value, using the linear regression which is that person's calibration curve. When the measurement is below FHFLEX, oxygen consumption is assumed to be the mean of the three calibration measurements made at rest. Energy expenditure is calculated using some value of heat production per litre of oxygen consumed, at a measured or assumed RER. For the time spent sleeping, energy expenditure is assumed to be equal to the BMR, which should be measured if at all possible or estimated from body weight and/or other physical characters using prediction equations. Energy expenditures for the three periods are then summed to give a 24-hour value for total energy expenditure. As with the use of activity diaries, a recording period of 3 to 7 days is recommended if the aim is to obtain some estimate of habitual energy expenditure.

Estimation of basal metabolic rate

Although the measurement of BMR is theoretically simple, it can be difficult to do. As originally defined by Boothby and Sandiford (1929), BMR is measured in the morning upon awakening, before any physical activity and 12 to 18 hours after the last meal. It is very easy to obtain elevated values in subjects who are unaccustomed to the methods and who may be restless or find the protocol stressful. When measured by experienced observers, long-term intraindividual variations in BMR have been shown to be small, even in subjects whose food intakes and activity patterns are not controlled (Shetty & Soares, 1988). Therefore, the BMR of an individual needs only to be measured once in the course of energy-expenditure measurements, unless changes in body size and composition suggest that BMR may have altered as a consequence. However, metabolic rate of women is known to fluctuate as a function of the menstrual cycle (Solomon, Kurzer & Calloway, 1982; Webb, 1986; Bisdee, James & Shaw, 1989; Meijer *et al.*, 1992), the sleeping metabolic rate during the post-ovulation phase being on average 7–8% greater than pre-ovulation rates. Therefore, it is important to standardise the timing of measurement of BMR or RMR of women, such that all measurements are made at similar stages in the menstrual cycle, preferably when subjects are pre-ovulatory.

RMR is a similar measurement to BMR, except that it is collected under less stringent conditions and generally gives less accurate results. Protocols for measuring RMR vary enormously and include differences in: (1) the restrictions placed on the subject in the 12-hour period before measurement; (2) the position of the subject during measurement; (3) the duration of the rest period prior to measurement; and (4) the duration of the measurement period. Furthermore, under field conditions, some researchers have their subjects transported to the laboratory for measurement after spending the night in their own homes, while others spend the night in the setting in which the measurement is to be made. Such variation in protocol leads to greater between-study variation in mean values obtained than with BMR; furthermore, protocols that are permissive in terms of activity prior to measurement and/or with a short pre-measurement rest period are likely to give higher RMR values than less-permissive protocols. Finally, measurement of RMR, with the generally less-stringent measurement protocols than for BMR measurement, will, on average, give higher values than BMR. Differences in protocol make it difficult to compare RMRs across populations; however, within-group comparisons are quite valid if the same methodology is used throughout.

BMR is the more accurate measurement; the importance of direct measurement lies in the fact that a number of populations have been shown to have different values for mean BMR per unit of body weight from Europeans. Notably, Schofield (1985) showed that Indian subjects had lower BMRs than Europeans and North Americans, while Shetty and Soares (1985), McNeill *et al.* (1987) and Piers and Shetty (1993) found that BMRs in Indian subjects were 7–12% lower than predicted from body weight using equations derived from predominantly European data (Schofield, 1985). In an analysis of BMR data collected in tropical populations, Henry and Rees (1991) showed that, on the basis of body weight, the Schofield (1985) equations over-predict BMR in a number of tropical populations. In males, they over-predict BMR by 22% (Sri Lanka), 9% (Philippines and Malaysia), 7% (non-European Hawaians), 6% (Japan) and 5% (Java), while in Sri Lankan females, they overpredict by 12%. Such differences have been attributed to either: (1) indeterminate 'ethnic factors', which when examined in different populations sharing the same environment show no effect (de Boer *et al.*, 1988); (2) tropical effects, which, although difficult to explain, have been shown in some individuals (Mason & Jacob, 1972) but not others (Hayter & Henry, 1993); and (3) subtle differences in body composition, including size of organs with very high metabolic activity, which cannot yet be measured (Elia, 1992). The balance of evidence is in favour of the third explanation (Elia, 1992; Shetty, 1993), although it has been suggested that the 'tropical effect' may be a function of marginal undernutrition in subjects where the down-regulating effects of negative energy balance on BMR (Waterlow, 1990; Ulijaszek, 1995) might be operating (Hayter & Henry, 1993). However, regardless of the reason for the effect, clear differences in BMR have been demonstrated between Europeans and some non-European populations, and unless it is clear that BMR either follows the European pattern or is different to some quantifiable extent, there is a need to measure BMR.

Prediction of BMR

It is not always possible to measure BMR, however, and there are a number equations which allow BMR to be predicted, although there are problems associated with their use. The Harris and Benedict (1919) equations are widely used in the USA and have been shown to over-predict the BMR of healthy Americans by between 10 and 15% (Daly *et al.*, 1985), while the Schofield (1985) equations adopted for international use (FAO/WHO/UNU, 1985) over-predict measured BMR in a number of non-western

Table 3.9. *Standard errors for prediction (coefficients of variation) of
BMR from body weight using the Schofield (1985) equations*

Age group (years)	Standard error of prediction (%) Males, group size			Females, group size		
	1	10	100	1	10	100
>3	19.4	6.3	2.5	16.0	5.2	2.1
3–9.9	6.8	2.2	0.8	7.6	2.4	0.9
10–17.9	7.5	2.4	0.8	9.3	3.0	1.0
18–29.9	9.3	2.9	0.9	9.3	3.0	1.0
30–59.9	10.4	3.3	1.1	8.3	2.6	0.9
60+	12.4	4.2	2.1	9.4	3.3	1.8

Source: From Schofield (1985).

populations (Henry & Rees, 1991). There are also equations for predicting
BMR in tropical populations (Henry & Rees, 1991) and Indian males
(Soares, Francis & Shetty, 1993). Care must be taken in choosing the most
appropriate equations for the population being studied; when in doubt, the
default option is to use the Schofield (1985) equations. It should also be
noted that the errors associated with the use of these equations for
predicting group means becomes smaller the larger the sample size is. Table
3.9 gives standard errors for prediction using the Schofield (1985)
equations, expressed as a percentage of mean BMR or coefficients of
variation. Clearly, the errors surrounding the prediction of any individual
BMR are extremely large, becoming smaller (less than 5%) with groups of
ten or more in all age groups except the youngest.

Doubly labelled water
The doubly labelled water method was developed by Lifson (Lifson &
McClintock, 1966) for use in animal studies and was adapted for use in
studies of humans by Schoeller and colleagues (Schoeller *et al.*, 1980).
Subjects are given an oral dose of water labelled with the stable,
non-radioactive isotopes of hydrogen and oxygen. The deuterium labels
the body's water pool and the oxygen-18 labels both the water and
bicarbonate pools which are in rapid equilibrium. The two isotopes are
excreted from the body at different rates. Deuterium leaves as water, but
oxygen-18 leaves both as water and as carbon dioxide. The difference in
loss rates between deuterium and oxygen-18 is proportional to carbon
dioxide production, from which energy expenditure can be derived.

Further details of the method are given by various sources (Schoeller *et al.*, 1980; Schoeller & van Santen, 1982; Schoeller, 1983; Prentice, 1988, 1990a).

The major advantage of this method is that the field aspects are extremely simple, and all the complexities are kept in the laboratory. The subjects are asked to drink an accurately weighed amount of the labelled water after an overnight fast. They continue to fast for an additional 4 hours equilibration period, after which they can return to their habitual routine. The disappearance rates of the isotopes are measured by mass spectrometric analysis of serial urine (Prentice, 1990a) or saliva (Stein *et al.*, 1987) samples. Two versions of the method exist, regardless of whether urine or saliva samples are used; the first requires the collection of a sample on each of the measurement days (Coward & Prentice, 1985), while the second only requires single samples to be taken at the beginning and end of the measurement period (Schoeller, 1984; Schoeller & Taylor, 1987). Validation studies show both versions of the method to give only small discrepancies from whole-body calorimetry values, for grouped data, for adult subjects exhibiting low to moderate levels of physical activity (James *et al.*, 1988b). The multi-point method offers a precision which ranges between 1 and 5% compared with whole-body calorimetry, while that for the two-point method ranges between 2 and 8%, with no significant bias between the two methods (Nagy, 1990).

Although this method is attractive, it is too expensive to be widely used. Furthermore, since the method measures carbon dioxide production, knowledge of the macronutrient composition of the diet is needed to convert carbon dioxide production into energy expenditure. This is similar to estimations of energy expenditure from oxygen consumption; in the absence of dietary data which can give some estimate of RER, a value of 0.85 gives an approximation of the heat equivalent of carbon dioxide production when consuming a mixed diet. Errors owing to uncertainty in macronutrient intake are typically less than 3% (Black, Prentice & Coward, 1986). In addition, the method has been shown to give unreliable estimates in subjects whose energy expenditure undergoes a profound change during the measurement period, or who sweat profusely, causing high total body-water turnover. Both situations might be encountered in groups engaged in subsistence production, where people may engage in hard physical work on one day but not the next, and where sweating might be caused by heavy physical activity or hot dry ambient conditions, or both. Therefore, it is important to keep some crude but independent assessment of physical activity, as well as an estimate of temperature and humidity while carrying out measurements, to aid the identification of anomalous results. In addition, the doubly labelled water method appears to be inappropriate for energy-expenditure estimation of lactating women, largely because of high water turnover in this group (Lovelady *et al.*, 1993).

Modifications to the standard protocols have been proposed by various authors to allow the measurement of energy expenditure by this method in difficult circumstances, notably for infants (Roberts *et al.*, 1988), and during heavy exercise and high activity levels (Westerterp *et al.*, 1986, 1988).

Physical work capacity ($\dot{V}O_2$max)

The measurement of oxygen consumption associated with work has been extensively and critically reviewed (Passmore & Durnin, 1955; Banister & Brown, 1968; Astrand & Rodahl, 1986; Nelms, 1982). It is assumed that heart rate has a well-defined relationship to work and oxygen consumption, and the aim of any test in which maximal aerobic power is measured is to determine the capacity of an individual's body to perform aerobic work. This can be done either directly, by exercising a subject to exhaustion by a stepwise protocol involving the use of a bicycle ergometer, treadmill or standard step test and measuring oxygen consumption when close to exhaustion; or indirectly, by predicting maximal oxygen consumption by extrapolating the relationship between heart rate and oxygen to an hypothetical, or population, maximal heart rate (Fig. 3.4).

Although it is preferable to obtain a direct measure of $\dot{V}O_2$max, in

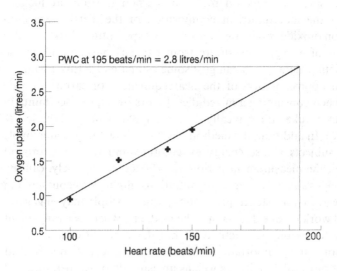

Figure 3.4.   Prediction of maximal oxygen uptake from an extrapolation to the subject's presumed maximal heart rate (in this case, 195 beats/minute). PWC, physical work capacity defined as the volume of oxygen consumed per minute at the maximum heart rate ($\dot{V}O_2$max (195)).

practice this is rarely done among rural subsistence-oriented populations. This is because tests to exhaustion may alienate subjects from taking part in further study, or subjects may be unwilling to push themselves to the limit. Furthermore, tests to exhaustion are potentially dangerous, particularly for individuals with undiagnosed heart conditions and for older adults. Two types of test are recommended for the field, the bicycle ergometer and the step test. In practice, the step test is most often used since it can be constructed in the field, is inexpensive and, in many contexts, subjects are more likely to be familiar with stepping than with riding a bicycle (Nydon & Thomas, 1989).

In submaximal tests, extrapolation to an hypothetical maximal heart rate is a source of unquantifiable error, for two reasons: (1) the predicted heart rate is taken to be a population mean to which the relationship between heart rate and oxygen consumption is extrapolated for all individuals, regardless of the knowledge that maximal heart rate varies from subject to subject in all populations; and (2) predicted maximal heart rate derived from values obtained for western populations may not apply to populations in the tropics. Furthermore, predicted maximal heart rates for children may differ between western, tropical, and high-altitude populations. Despite these drawbacks, estimates of $\dot{V}O_2$max can give some measure of the ability to perform hard work and of endurance. When matched with the energy cost of performing important subsistance tasks, $\dot{V}O_2$max can be used to determine the extent to which individuals within a group might be physically stressed by hard work. Such quantification has allowed biological anthropologists to study subsistence ecology in new ways (Spurr, 1988a,b). Performance effort is the proportion of maximal aerobic power expended in the performance of any given subsistence task. The greater the proportion of an individual's $\dot{V}O_2$max, the greater the performance effort required. When used in association with activity diaries across the year, it is a quantitative way of determining whether work in subsistence is more stressful at one time of year than at another. It can also be used to determine whether women work harder than men.

Energy balance

In individuals, differences between energy intake and expenditure over periods of more than a month can be inferred by estimates of energy balance. These are simpler to do than either intake or expenditure and involve anthropometric measurements such as weight, arm circumference and skinfold thicknesses at various sites. These may be used either independently of, or in association with, intake and/or expenditure methods.

Changes in anthropometric measurements that are taken to be indicative of changes in energy balance are often quite small, and it is important to have some estimate of measurement error to aid the interpretation of such data. The accuracy of collected anthropometric data can be evaluated by using two error estimates: the technical error of measurement (TEM) and coefficient of reliability (R) (Mueller & Martorell, 1988; Frisancho, 1990, 1993). A detailed account of the measurement and use of measurement error estimates is given in Ulijaszek & Lourie (1994), who give reference values for upper limits for TEM for arm circumference, and triceps and subscapular skinfolds.

Body composition measures are useful for estimating energy balance in adults, particularly when the same individuals can be compared across time. The most common method is estimation of body density and subsequently body fat from regression equations of skinfold thicknesses aginst body density (Lohman, 1988). Such equations have been derived from studies in which the skinfolds of a large number of subjects have been correlated with their body density as estimated by a reference method, such as densitometry or isotope dilution. Although a large number of prediction equations have been developed in this way (Sloan, 1967; Durnin & Womersley, 1974; Jackson & Pollock, 1974; Jackson *et al.*, 1980), only a few have been validated across different populations. The equations of Durnin and Womersley (1974) have been more widely validated than any other (Ulijaszek, 1992a) but do not work for older New Guinea adults (Norgan, Ferro-Luzzi & Durnin, 1982), Eskimos (Shephard *et al.*, 1973), or Inuit (Rode & Shephard, 1994). A problem exists in the use of prediction equations based on the measurement of skinfolds in relation to middle-aged and elderly people who have experienced a significant amount of bone loss; on average, as bone mass is lost, body fatness is over-predicted by an amount that is correlated with the degree of mineral loss (Mazess *et al.*, 1984). Williams *et al.* (1992) have developed a multiple-component approach which corrects for this bias in North American subjects. In addition, predictive equations derived from measurements of non-pregnant, non-lactating women have been found to be appropriate for lactating women, despite differences in patterns of fat deposition (Butte *et al.*, 1985).

Bioelectrical impedance and resistance

The bioelectrical impedance (BEI) and bioelectrical resistance methods depend on the differences in electrical conductivity of fat-free mass (FFM) and fat and, because of the ease of use and the portability of the equipment, have been promoted as potential field methods for estimating body composition. These techniques measure the impedance of a weak electrical current ($800\,\mu A$; $50\,kHz$) passed between the right ankle and right wrist of an individual. The impedance is proportional to the length of the conductor and indirectly proportional to the cross-sectional area. The length of the conductor is usually a function of the height of the subject. A number of predictive equations have been developed (Lukaski *et al.*, 1985; Kushner & Schoeller, 1986; Guo *et al.*, 1987; Khaled *et al.*, 1988; Segal *et al.*, 1988) by relating BEI measures to a reference method of estimating body composition, either isotope dilution or densitometry. However, the accuracy of these equations in measuring FFM and fat in different populations has yet to be determined, although prediction equations have been developed specifically for normal (Tanaka *et al.*, 1990), and obese (Tanaka *et al.*, 1992) Japanese women.

Caution is urged in the use of this method, since it has been found that two equations which are used with commercially available instruments under-estimate and over-estimate FFM as determined from an isotope-dilution method by an average of 4.7% and 8.1%, respectively (Pullicino *et al.*, 1990). In the same study, prediction equations using skinfold thicknesses, weight/height and BMI gave under-predictions of only 2.6%, 4.1% and 2.8%, respectively. In another study, FFM of Pima Indians estimated from bioelectrical resistance was on average 5.3% lower than values obtained using the technique of hydrostatic weighing (Rising *et al.*, 1991).

Extensive cross-validation of equations has been carried out in European populations (Lukaski *et al.*, 1986; Hodgdon & Fitzgerald, 1987; Segal *et al.*, 1988; Guo, Roche & Houtkooper, 1989; Houtkooper *et al.*, 1989, 1992; Fjeld, Freundt-Thurne & Schoeller, 1990; Heitmann, 1990; Kushner *et al.*, 1990; Deurenberg *et al.*, 1991; Danford *et al.*, 1992). Most studies indicate that equations are generalisable. However, although the method has been validated for European populations (Heitmann, 1994), studies are needed to validate or modify the method for non-European populations. One modification, the use of multiple frequency bioelectric impedance, has been shown to increase reliability of measurement in North American subjects (Chumlea *et al.*, 1994) and can offer further discrimination in body-composition measurement, since low-frequency impedance can be used to estimate extracellular fluid, and high-frequency impedance can be used to assess total body water (Thomasset, 1963). This has been confirmed by Segal *et al.* (1991) and van Loan and Mayclin (1992). Improvements in this method, and more extensive validation, may make it a promising tool for use in non-European populations in the future.

Dual-energy X-ray absorptiometry (DEXA)

This method was developed for the precise measurement of bone mineral (Mazess *et al.*, 1981) but was subsequently adapted for soft tissue measurement as well (Mazess *et al.*, 1990). The method relies on the energy dependence of the attenuation coefficients for photon absorption of bone mineral, which contains calcium, an element of high atomic number, and soft tissue, which has large quantities of the low atomic number elements carbon, hydrogen and oxygen (Mazess *et al.*, 1990). The precision of DEXA in bone measurement is excellent, with a coefficient of variation of about 1% (Mazess *et al.*, 1990). DEXA measurements of three compartments, bone, bone-free lean mass and fat, when summed have been able to predict body weight accurately. However, a comparison of body composition estimates in small piglets using DEXA and carcass analysis showed DEXA to under-predict total body weight by only 0.2%, but to over-predict fat mass by 114%, and under-predict lean mass by 6% (Brunton *et al.*, 1992). The over-prediction of fat mass by 93 g was compensated by an under-prediction of non-bone lean mass of 85 g. This method is probably acceptable as a reference method for bone-tissue estimation but not yet for non-calcified tissues (Roubenoff *et al.*, 1993).

Indices and ratios

There are a number of derived indices and ratios that are helpful in human energetics. These include: (1) output–input ratio; (2) body mass index (BMI); and (3) physical activity level (PAL).

The output–input ratio is a measure of ecological efficiency, in which the food energy produced is divided by the total human energy expended in that production (Ellen, 1982). Patterns of energy expenditure in a community or population are a function of the type and extent of local energy acquisition, exchange relationships with other groups and other activities that are important for the maintenance of group function, including individual biological factors as well as group economic maintenance and reproduction (Nydon & Thomas, 1989). The output–input ratio can be obtained by applying appropriate energetics methods and can give some idea of the extent to which a group, community or population are successful in the food quest. For groups involved in traditional subsistence, values below about 8 are incompatable with long-term survival, while values in excess of 8 reveal the extent to which the group is able to generate surpluses and, thence, be able to buffer itself against dietary shortfalls caused, for example, by environmental unpredictability.

BMI can be used as a measure of current nutritional status in adults and is weight in kilograms divided by the square of height, in metres. For adults, values of 25, 30, and 40 have been designated cut-offs for grades I, II and III obesity in the UK by Garrow (1981), while values of 18.5, 17 and 16 have been designated cut-offs for mild, moderate and severe chronic energy deficiency internationally by Ferro-Luzzi *et al.* (1992). Although crude, this measure gives an easily calculated measure of energy nutritional status of adults.

PAL is a rough measure of how physically active people are. This is obtained by dividing total daily energy expenditure by BMR. It can also be divided by RMR, if this is the variable measured, but it should be acknowledged that this will give lower values for PAL than would be the case if BMR were applied. Values proposed as indicative of light, moderate and heavy work schedules respectively are: light: 1.55 (men), 1.56 (women); moderate: 1.78 (men), 1.64 (women); and heavy: 2.10 (men), 1.82 (women) (FAO/WHO/UNU, 1985). The PAL value which is considered to be the minimum habitual level compatable with life is 1.27 (FAO/WHO/UNU, 1985).

Summary

Considerable advances have been made in most aspects of energetics methodology in recent times, above all in energy expenditure measurement. By partitioning energy expenditure of adults between maintenance and activity and measuring body composition, it is possible to examine possible specific adaptations to low-energy availability, notably the level and efficiency of physical activity and down-regulation of basal metabolism. It is also possible to estimate different components of energy expenditure by infants and children.

Measures of energy intake are beset with problems, the most important being one of under-reporting or under-eating at the time of study. Of studies in which energy intake and expenditure have been measured in parallel in the same subjects, about half show mean levels of intake that are much lower than expenditure, suggesting under-estimates of intake. A huge literature has been generated by researchers addressing problems associated with dietary intake measures. Techniques have been developed to assess the reliability and accuracy of data, which help in the analysis and interpretation of energy-intake data.

Differences between energy intake and expenditure over periods of more than a month can be inferred by estimates of energy balance. These involve some measure of body composition, and a variety of techniques, newer and older, are available. Evaluation of some of the newer techniques in different populations and under different conditions is in process, but improved body composition technologies promise to provide the next breakthrough in energetics methodology.

4   *Modelling*

Modelling has been important in the development of ecological energetics, and although the focus has shifted away from this approach in recent times, modelling can still furnish insight into human adaptive processes. Models are partial representations of reality, the components of which are thought to be important in the system under investigation. Since the assumptions and relationships between components reflect current understanding, and may change with time, models are simply heuristic tools whereby complex phenomena can be examined, and from which testable hypotheses can be generated (Thomas *et al.*, 1982). As such, they vary according to the purposes of the investigator, the focus of investigation and the part of the ecosystem of particular interest. Therefore, models vary in the relative degrees of reality, generality and precision they contain, and no two models are alike. Briefly, precision is the degree to which data collection manages to capture all the important features of the ecosystem under study; reality is the extent to which the information, once analysed, is biologically and culturally plausible; generality is the extent to which the findings from one study can be compared with findings from other studies. General models allow comparisons between ecosystems but, since no two ecosystems are the same, must involve simplification. This can lead to loss of information, which may be important in the understanding of that ecosystem. At the other extreme, collecting data to a high degree of precision allows detailed understanding of the ecosystem under study but makes comparison with other ecosystems difficult.

Modelling has been used to: (1) attempt to understand and organise causal relationships between different components of human ecosystems; (2) predict the consequences of change in one part of a system on other parts; (3) compare different systems, and their energetic efficiencies; and (4) identify areas requiring greater or more intensive investigation. Energy economic relationships can reveal aspects of exchange that could be of adaptive significance in communities which rely minimally on cash-based exchange, or in parts of the system that cash cannot represent. For example, biological differences can be found in most parts of the world between populations of higher and lower economic status. Being poorer, as measured by purchasable material resources or cash income, means having limited access to resources and living in poor environments, often with poor health and slow growth of children. Energy can quantify non-monetary relationships which may have similar biological correlates. For example, taller, heavier men may be physically fitter, work harder and produce more food for their families than smaller men. As a consequence, their families may be bigger and suffer less disease than the smaller man's family. Furthermore, the wife of the taller man may have larger babies, be more successful in breastfeeding and have more viable offspring. Such relationships could not be revealed if money were the currency of exchange being used. Energy modelling is similar to economic modelling, differing in the choice of currency and the type of question asked. In this chapter, the principles of energy modelling are outlined.

Human ecological modelling

Human ecology is the study of the interrelations that exist between individuals, populations, and the ecosystems of which they are a part (Ulijaszek & Strickland, 1993b). Since most of human ecology involves complex interactions, theoretical positions are often presented as models or mechanisms in a form that can be tested with data. A model can, therefore, be considered as a type of theoretical framework, while modelling is an analytical process in which biological responses to a range of plausible environmental and ecological circumstances can be explored.

Biological anthropologists have used modelling procedures to: (1) examine the relationships between social and biological processes in communities; (2) compare measures of ecological success of groups with different subsistence practices; (3) predict outcomes of change; and (4) generate new hypotheses. Generally, the data used in the creation or testing of models can be of various sorts. Time use, material acquisition and material exchange can all be used to measure ecological success. However, these variables are not interchangeable across different sectors of the subsistence system. Energy, as an interchangeable currency, allows comparison and analysis of disparate aspects of human ecological systems, including returns from hunting, gathering, herding or agriculture, in relation to effort exerted and expended in subsistence activities and the investment of time and energy in the raising of offspring. The advantage of energy modelling is that there is a common currency which flows through most sectors of the subsistence system, allowing the processes of production and reproduction within any community to be compared or related directly, once both have been translated into energy units.

Although the use of energy as an important factor in anthropology begins with the work of White (1949), energy modelling has a more recent history. As a technique, modelling was closely associated with the development of systems theory (von Bertalanffy, 1968; Buckley, 1967), an approach to problem solving. Definitions of 'system' include the following: (1) 'a set of objects together with relationships between those objects and their attributes' (Hall & Fagan, quoted in Langton, 1973); (2) any set of specified variables in which a change in the value of one of the variables will result in a change in the value of at least one other (Rappaport, 1968); (3) a set of units with relationships among them; the word 'set' implies that the units have common properties; the state of each unit is constrained by, conditioned by or dependent upon the state of other units (Millar, 1965). Following this, Langton (1973) concluded that units in any system are causal, functional or normative, the 'set' being organised through the interrelationships between units, and the units exist as a whole that is greater than the sum of its parts. Thus, ecosystemic modelling explicitly analyses biological, environmental and behavioural traits as part of a single system. Systems analysis in this context involves (1) defining the goals and objectives of the study or analysis; (2) simplifying the system to define measureable aspects of the important variables relevant to (1); (3) representing the relationship between the variables in a way convenient for analysis; and (4) establishing the system boundary, in relation to the question in hand.

Early models stressed the importance of the homeostatic nature of human societies (Rappaport, 1968, 1971). Loops of circular causality, or feedback, operate between units of the system to cause mutually adaptive changes in response to changes in the environment (Ellen, 1982). However, the notion of homeostatic processes outside of the physiological context must be qualified by the understanding that feedback may serve the adaptive process, but that the process of adaptation does not result in a return to the previous, unperturbed state. Rather, all relationships between units may changed to a varying degree, and the local, human-made environment is also likely to be modified in the process. Therefore, the preservation of equilibrium can at best only be approximate. For example, one type of change in response to environmental stress in rural third-world communities involves the modification of subsistence behaviours and practices. Decisions to plant different types of crop have implications for energetic relationships between systemic units such as work organisation and output, the type and amount of food acquired, the distribution of that food between males and females, adults and children, and the functional implications of the plane of energy nutritional status maintained as a consequence. In addition, changes in subsistence practices, however small, will have an impact on the environment, changing it in either a positive or negative manner, but changing it nonetheless. Following from this, environmental change driven by human action leads to new environmental stresses which require further modification of behaviours and practices. The relationships between human action and environmental change are, therefore, more homeoretic than homeostatic and are akin to 'tracking a moving target'. It also has consequences for human biology, which may be homeostatic with respect to short-term physiological processes, but not with respect to longer-term human population biological characteristics.

Although energetics models require systems closure to work, it is often difficult to determine where the boundaries should be placed. In real life, all systems are open, since even the remotest group has relationships with other populations. For example, local societies are rarely closed reproductive units and are, therefore, demographically part of larger groupings. Furthermore, societies generating surpluses or producing valued goods are able to trade with other groups, either adjacent or more distant. The process of modernisation has brought most of the world's groups and populations into the cash economy at least to some extent, allowing them to trade globally. For example, island populations in Papua New Guinea and the Solomon Islands may be physically and geographically isolated, but they still engage in cash-cropping and are able to buy goods, such as tinned fish, which has come from Japan, or bush knives, which come from Britain. The availability and use of both can change the energetic relationships in a community. In the latter case, improved efficiency of some aspects of agricultural work are made possible by the availability of steel tools. Once boundaries of the model have been set, data collection and preliminary analysis may determine that the boundaries should be modified to allow phenomena emerging as a consequence of these processes to be considered. Once analytic closure has been achieved, analysis can proceed through the construction of models linking variables as part of causal processes. Adaptive processes occur in systems only when stresses push a parameter beyond its range of stability (Millar, 1965). Although it is difficult to define what these limits might be for a number of parameters, it is possible to determine the functional implications of energetic stress on physiological systems, at least.

Types of model
Models can be descriptive, analytical or predictive. By far the most common use of modelling has been descriptive, as a convenient way of summarising energy relationships in a community and, by use of output–input ratios, determining the likelihood of energetic stress. Analytical

models are more difficult to use but can determine the types of energetic stress encountered in a community, and which parts of the community are likely to be affected. Predictive modelling has aroused interest as an heuristic tool in examining the population effects of possible changing social, environmental or biological situations (Thomas *et al.*, 1982). The usefulness of all types of model is limited by the extensiveness and accuracy of the data collected.

Models of hunter–gatherer foraging strategies have largely concentrated on time spent for hunting gains, giving some idea of the optimisation of various outputs and returns. Very few studies of hunter–gatherers have examined energetics. Notable is the early study by Lee (1968) of the !Kung bushmen of the Kalahari desert (the major conclusions of this study are given in Chapter 1) and the studies of the Ache of Paraguay (Hawkes, Hill & O'Connell, 1982), which show that the energy returns of a number of subsistence activities to be very high in relation to input. Energy modelling of hunter–horticulturalist and swidden agriculturalist groups has been more common (Rappaport, 1968; Morren, 1977) as has that of agricultural societies operating at various levels of technology (Bayliss-Smith, 1982b).

Descriptive models

Descriptive models are by far the simplest, and Figs. 4.1 and 4.2 give the basic features common to this type of model. Figure 4.1 shows the flow of energy in a generalised and simplified ecosystem. There are two types of food chain represented here: grazing and decay. In a grazing chain, energy accumulated by primary producers, mainly plants, is used by primary consumers (herbivores). These are then consumed by secondary consumers, or carnivores. There may also be tertiary consumers (larger carnivores that consume smaller ones), while omnivores are able to span several consumption levels. Humans, for example, are primary, secondary and tertiary consumers. The key given in Fig. 4.1 was first put forward by Odum (1971) and is the common convention used in descriptive energetics modelling. Figure 4.2 gives a schematic representation of energy pathways for a hypothetical human population. Thus, it can be seen that most aspects of human activity can be included in the descriptive model, as long as an energy value can be obtained for it. The relative importance of different components will vary from group to group, and some components prominent in one society may be completely absent in another. The important point is that this scheme is a starting point, based on data collected on aspects of the system considered to be important by the researchers. Figures 4.3 to 4.6 give descriptive energetics for three groups, the Tsembaga Maring of Papua New Guinea (Rappaport, 1968), the

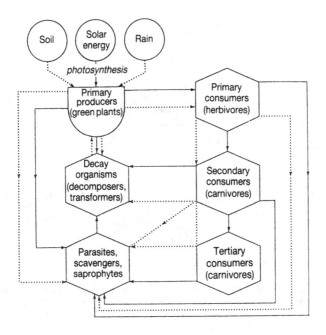

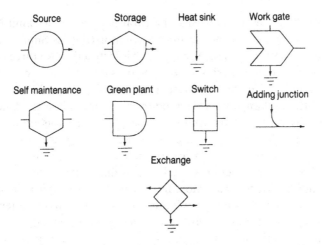

Figure 4.1. The flow of energy and of nutrient materials in a generalised and simplified ecosystem. Continuous lines represent energy flows and broken lines flows of nutrients. The key is to conventions derived from Odum (1971) and used in this and subsequent figures. (From Ellen, 1982.)

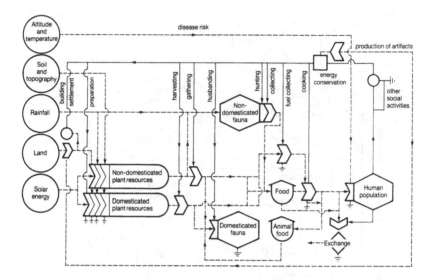

Figure 4.2.   Schematic diagram of energy and nutrient pathways for an hypothetical human population. Broken lines represent food chains and other inputs, while continuous lines represent pathways of energy expenditure and control. (From Ellen, 1982.)

Quechua of Peru (Thomas, 1976) and a rural population in Tamil Nadu, South India before and during the green revolution (Bayliss-Smith, 1982a).

Simple examination of these three systems is instructive. All three vary in the extent to which connectedness with the outside world is important for their functioning. The least connected are the Tsembaga Maring, who engage in little exchange with neighbouring groups, while the Quechua depend heavily on trade to sustain the ecosystem that was observed by Thomas (1976). The South Indian peasants were heavily engaged in trade with the larger economic world before the green revolution; after taking on a more technological agriculture, their involvement increased.

The Tsembaga Maring rely most heavily on the mechanical energy of human labour. Of a total population of 204, 127 were potentially productive, and these persons worked on average of 9.5 hours per week in food-producing activities. The greatest work input is into planting and weeding of sweet potato, taro and yam gardens, and herding of pigs. Of the harvested crops, about a quarter go to feed the domestic pig herd. The remaining three quarters supply 91% of the dietary energy intake; pigs supply about 8%, with hunted and gathered foods supplying 1% of energy intake. Overall, the Tsembaga Maring achieve an output–input ratio of 14.2 (Bayliss-Smith, 1982b), and, therefore, seem to generate a small

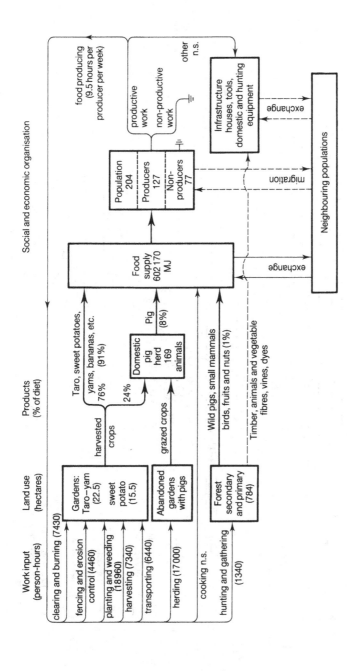

Figure 4.3. Energy inputs and outputs and material flows in the Tsembaga Maring agricultural system. Energy flows: ——, human work; —·—, food. ······, movement of goods and people; n.s., no statistics. (From Bayliss-Smith, 1982a.)

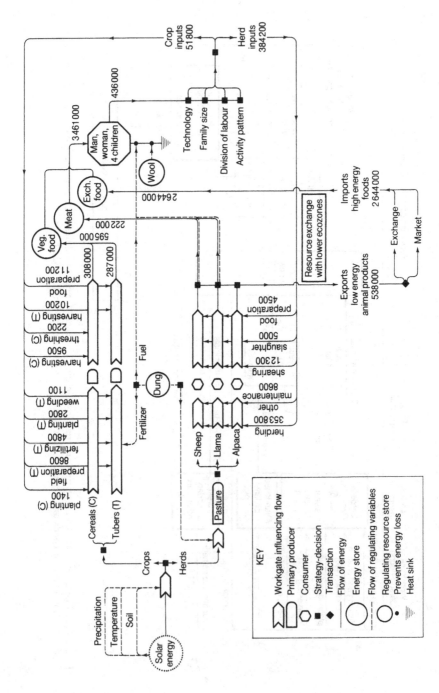

Figure 4.4. Annual energy flow (in kilocalories) through a typical Nunoan Quechua family (from Thomas, 1976).

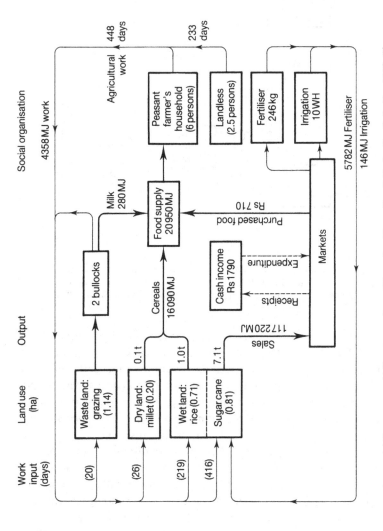

Figure 4.5. Energy and material flows on a middle peasant farm, Wangala village. Tamil Nadu, South India. 1955. Flows: —, work energy; —, food energy; – –, movement of money and goods. (From Bayliss-Smith, 1982a.)

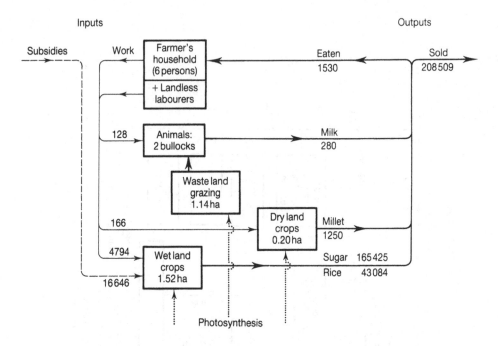

Figure 4.6.   Energy flows (in MJ) on a middle peasant farm, Wangala village, Tamil Nadu, South India, 1975–6. Energy flows (in MJ): —, work; —, food; – – –, fertiliser, irrigation, pesticide; ...., solar. (From Bayliss-Smith, 1982a.)

surplus of food energy with relatively small work inputs. The output–input ratios of different subsystems of the subsistence economy are: crop production, 20.9; pig keeping, 3.2; and hunting and gathering, 5.4 (Bayliss-Smith, 1982a). This indicates that crop production subsidises pig keeping (see Chapter 1) and that hunting and gathering, as practised in the 1960s, could not sustain this population. There are ritual aspects of the subsistence economy which are related to land-use and the carrying capacity of the land (Rappaport, 1968); these are discussed in Chapter 1.

The Quechua Indians live at high altitude in the Peruvian Andes, and practise agro-pastoralism. The herding of animals uses 88% of all human energy input into subsistence work. However, herding and agriculture are related, since the impoverished soils of the Peruvian Altiplano require plentiful fertiliser for any adequate and sustained agriculture. This fertiliser comes in the form of dung, from the sheep, alpaca and llamas that are herded. Thus, herded animals consume scrubland vegetation on marginal land and produce fertiliser which allows sustained cultivation to be practised.

Trade is of fundamental importance to this group of Quechua. Without trade, the output–input ratios of different subsets of the subsistence economy are: crop production, 11.5; and animal herding, 2.0. Although it might seem that crop production should sustain this group even in the absence of other subsistence activities, it would only supply about 70% of possible energy returns were herding also to be practised. This is despite the practice of herding having extremely low energetic returns. However, animal products command a respectable price on the open market, and the energy-exchange ratio of animal products for foods, predominantly cereal flours, is 4.9. Therefore, for every megajoule of meat, wool or animal skin sold, 4.9 MJ of wheat or maize flour can be bought. Without trade, the existing subsistence system would provide an overall output–input ratio of 3.1; with trade, the ratio is 7.9. This is low, but suggests that it might sustain the population, albeit at marginal energy nutritional status. This is borne out by measures of growth and development in this group. Poor growth relative to international references is one measure of nutritional status (Ulijaszek & Strickland, 1993b), and the Quechua of Nunoa have growth patterns which are below the 5th centile of NCHS references for stature and weight (National Center for Health Statistics, 1977); growth continues until the 22nd year, with final attained stature that is below the 5th centile of the NHANES III references for stature and weight (Frisancho, 1976, 1990).

The changing agricultural ecology of South India is illustrated by energy flow in an agricultural community in Wangala village, Tamil Nadu. In 1955, the majority of the community comprised peasant caste farmers (Epstein, 1962). Trade is important for this population and in 1955 took the form of the cash crop sugar cane. Although 61% of the total human work went into the cultivation of this crop, sales of which were substantial, purchased food only supplied 22% of total energy intake. Milk from bullocks only supplied 1% of total energy intake, but, as with the Quechua, manure from livestock was a vital source of fertiliser. In 1955, the overall output–input ratio was 13.0. By 1975–6, it was down to 9.7.

New rice varieties that were developed at the International Rice Research Institute in the Philippines became available in India in 1966. These improved varieties could increase yields, but they also required greater inputs. The impact of the green revolution created by the new agriculural technologies has been examined by Bayliss-Smith (1982a), and these are summarised in Table 4.1. In 1955, middle peasant farmers had a dual economy, with modernised sugar-cane production and traditional rice subsistence. By the mid 1970s, semi-industrial technology prevailed wherever there was irrigation. Overall yields had increased as a result of

Table 4.1. *Agricultural system of a middle peasant farmer in Wangala after the green revolution, assuming no change in farm size, land-use pattern, or the size of the farmer's household*

Land use	Input/output	Quantity (1975–6)	Energy value (MJ)	Change since 1955
Wet land Sugar cane				
(0.81 ha)	Human work	524 days	3354	+26%
	Fertiliser	585 kg	12 324	+138%
	Irrigation	5.4 kW/h	78	Nil
	Sugar cane	100.2 t	165 425	+41%
Rice (0.71 ha)	Human work	225 days	1440	+3%
	Fertiliser	166 kg	3526	Innovation
	Pesticide	7 kg	650	Innovation
	Irrigation	4.7 kW/h	68	Nil
	Paddy	4259 kg	43 084	+190%
Dry land Finger millet				
(0.20 ha)	Human work	26 days	166	Nil
	Millet	98 kg	1250	Nil
Waste land Bullock grazing				
(1.14 ha)	Human work	20 days	128	Nil
	Milk	(Regional average)	280	Nil
Farmer's household				
Composition		6 persons		Nil
Subsistence food supply (millet, milk)			1530	−90%
Purchased food supply (rice, meat, etc.)			19 420	+300%
Cash income				
Farmer (net household)[a]		Rs 2150		+20%
Labourer (per caput)[a]		Rs 91		−66%
Total farm energy input			21 734	+110%
Total farm energy output			210 039	+57%
Energy ratio		9.7		−25%
Energy yield (MJ/ha)			66 470	+57%

[a] Using Epstein's deflation index for Wangala 1955–70 of 2.85, further deflated by 28%, the increase in rice prices 1970–6, to an overall index of 3.65. Incomes of 1975–6 are, thus, expressed in terms of their 1955 value.

adopting high-yielding rice varieties. However, rice had become a cash crop. Yield had increased 57%, but the energy inputs required to obtain this level of output had increased 111%. However, incomes of those that could engage in growing high-yielding varieties of rice increased in real terms by 20%. Also, for some there was a shift from doing agricultural work, with casual labourers being bought in; more prosperous farmers then began to see agriculture more as a business than a way of life.

Analytical modelling

Analytical modelling is the process whereby aspects of the descriptive model are selected for more detailed examination, once it has been determined that a part of the system is under energetic stress. For example, the Quechua considered in Fig. 4.4 were shown descriptively to rely heavily upon trade for the maintenance of their ecosystem. Further examination shows that the time spent in herding exceeds the possible time budgets of adults alone, and that the size of animal herds used by the Quechua to obtain animal products for trade could only be maintained with the extensive use of child labour.

Indeed, the Quechua practice a high-fertility strategy in which large numbers of children are involved in the subsistence economy. Although many of these children die before reaching adulthood, a large number of them, on becoming adults, migrate in search of work in the towns and cities at lower altitudes. Thus, the children provide a source of labour that does not result in an increase in population, since most of the surplus adults arising out of a high-fertility strategy depart from the system permanently before reaching reproductive age (Thomas *et al.*, 1982). In this example, analytical modelling would involve developing the part of the descriptive model concerned with the energetics of reproduction, in particular fertility, as well as the energy costs of child rearing and the energetics of child labour.

Predictive modelling

Predictive modelling uses original data and either the descriptive or analytical model to examine changes to one or several outcome variables when some conditions in the human ecosystem change. One example of possible change is a reduction in out-migration from the Andean altiplano by young Quechua adults. The problem of population expansion among the Quechua is exported to the towns. It might be expected that, at some stage, increasing population size in urban areas might lead to a deteriorating physical environment and quality of life, and concomitantly lower expectations of success by potential migrants. Therefore, at some time in the future, the perceived benefits of migration might be lower than the

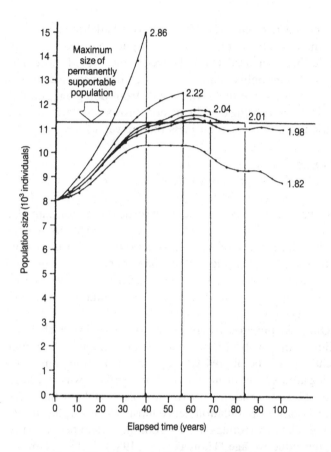

Figure 4.7.   The effect of instant decreases in mortality and birth rates on Quechua population size. Birth rate decreased from 6.67 children per female per lifetime to levels indicated on the graph (from Thomas *et al.*, 1982).

understood quality of life in the altiplano, and, rather than migrating, people might stay in the village. At this point, the rural population would increase. With some knowledge of the extent of urban stress and the extent to which urban populations can increase before perceived opportunities to potential migrants recede, possible effects of reduced migration could be modelled predictively. One example of such predictive modelling is given in Fig. 4.7, where the effects of reduced child mortality on population size are examined. This shows that unless fertility levels drop dramatically and immediately, then the population will rise to unsustainable levels within a generation. Thus, not only does this Quechua human ecosystem depend on high out-migration of young adults to maintain existing population levels, but high child mortality levels also play a very important part.

Modelling of hunter–gatherer subsistence

In the study of hunter–gatherer societies, the use of optimisation procedures have been used, following the postulates of synthetic evolutionary theory. Pianka (1988) states that 'natural selection and competition are inevitable outgrowths of heritable reproduction in a finite environment'. Thus, in the realm of human ecology, direct competition for resources within and between groups gives advantages to subgroups or groups that have efficient techniques of acquiring energy and nutrients that can be turned into offspring or used to avoid predators. In this way, the more efficient pass on more copies of their genes, and there is a relative increase in fitness for the group as a whole. Energy has become the nutrient of choice, since the energy drive is the primary one in most societies, the meeting of energy needs being the most fundamental in nutritional terms. Furthermore, energy gains can be perceived in relation to effort, which can be expressed in units of energy. So the optimisation procedure involves an examination of the efficiencies of energy capture.

Many foraging strategy models are based on geometric representations of the relationships of foragers to resources, with solutions found by graphs or by differential calculus (Cody, 1974). Once the currency is chosen, an appropriate cost–benefit function must be adopted, and the function solved for an optimum. Although the most fundamental biological property of an organism is reproductive fitness, it is rarely feasible to measure this. Instead, it is usually assumed that fitness varies directly with the rate of net energy capture that can be achieved while foraging (Pyke, Pulliam & Charnov, 1977). Although a number of factors can intervene to influence this relationship (Schoener, 1971), in general it holds true. Table 4.2 lists the optimality assumptions possible for a range of models which could be energy limited.

There are five main categories of optimality model that can be used to examine hunter–gatherer societies. In the diet-breadth model, the strategic goal is to determine the optimal set of resource types to exploit, optimality being measured by the return per unit of handling time for each type of resource. Conversion of units of resources in energy units allows direct comparison of the efficiencies of use of different resources. Adding these up allows the overall return to be estimated.

In the patch-choice model, the aim is to determine the optimal array of habitats to exploit, optimality being determined by the best set of patches to visit. This is estimated by measuring the average rate of resource returns in the different types of patch and comparing them. An efficiency ranking of patch types and habitat richness can then be created.

The time-allocation model aims to examine the optimal pattern of time

Table 4.2. *Optimal foraging models: cost–benefit criteria and major constraining variables*

Category	Strategic goal	Domain of choice	Cost–benefit criteria	Major constraints
Diet breadth	Optimal set of resource types to harvest	Which types to harvest	Return per unit handling time for each type; overall foraging time (including search time)	Search and pursuit abilities of forager, encounter rates with high-ranked types
Patch choice	Optimal array of habitats to exploit	Which set of patches to visit	Average rate of return with patch types and average over all patches (including travel time between patches)	Efficiency ranking of patch types, habitat richness, travel time between patches
Time allocation	Optimal pattern of time allocated to alternatives (patches etc.)	Time spent foraging in each alternative	Marginal return rate for each alternative, average return rate for entire set	Resource richness, depletion rates for each alternative
Foraging group size	Formation of optimalised groups for foraging	Size of groups to join for foraging under specified conditions	Average per capita return rate at each group size, marginal cost for group members	Return-rate curves for each group size under each condition possibilities for group formation, rules governing division of harvest
Settlement pattern	Optimal location of home base for foraging efficiency	Settlement location of each foraging unit (individual or family)	Mean travel costs and/or search costs per unit harvest	Spatiotemporal dispersion and predictability of major resources, effects of cooperation and competition

Source: Modified from Smith (1983).

allocated to different foraging alternatives, by comparing the marginal return rate for each alternative, in a given group. That is, the change in resource richness with the length of time spent foraging on a particular patch. Thus, the resource use can be rationalised in terms of environmental depletion.

In examining foraging group size, the goal is to determine the optimal size of group for foraging. This is estimated from measures of average per capita rates of return at each group size and by modelling the marginal costs or gains to the group and to new joiners, when the group size is either larger or smaller than that observed.

The settlement-pattern model considers the optimal location of the home base or camp for foraging efficiency. In this case, the settlement location of each foraging unit, be it individual, family or band is related to the travelling and searching costs per unit of resource harvested.

These five model types allow the examination of different aspects of hunter–gatherer foraging strategy. The use of more than one type of model provides more information than simply the sum of the individual models, and the understanding gained from one model may provide added insight into the optimality of another model. For example, application of the patch-choice model allows examination of the efficiency ranking of various patches used, while application of the time-allocation model allows the use of these patches to be related to their depletion rates. Thus, what may appear to be illogical by the application of one model may be shown to be perfectly logical by the application of the other. If several bands or groups are compared in this way, it is possible to examine the different rationalities behind the variety of foraging strategies seen in real life.

Summary

Energetics modelling can provide insight into human adaptive processes in that it can help to organise causal relationships between different human ecosystemic components. Furthermore, predictive modelling can be used to examine the consequences of change, as well as allowing comparison of the efficiencies of different subsistence systems.

Energetics models can be descriptive, analytical or predictive. By far the most common use of modelling has been descriptive, as a convenient way of summarising energy relationships in a community and, by use of output–input ratios, determining the likelihood of energetic stress. Models of hunter–gatherer foraging strategies have largely concentrated on time spent for hunting gains, giving some idea of the optimisation of various outputs and returns. Descriptive modelling of energy flow has been carried out for a number of populations. In this chapter, energy flow in three groups is described. These are the Tsembaga Maring of Papua New Guinea (Rappaport, 1968), a group of horticulturalists who engage in pig rearing; the Quechua of Peru (Thomas, 1976) who are primarily animal herders; and a rural population of rice cultivators in Tamil Nadu, South India, before and during the green revolution (Bayliss-Smith, 1982a). These examples show ways in which patterns of subsistence and the level of trade with other groups have profound influence on human population biology.

Energetics and anthropology

5   Reproductive performance

Energetics is important in different aspects of reproductive performance, including the onset of menses, fecundity, pregnancy outcome and lactational performance. Ultimately, the survival to reproductive age of as many children as possible is the most direct measure of reproductive performance, with all other factors being proximate (Ulijaszek, 1993b). Total fertility rate (TFR) is a commonly used measure of reproductive performance and is the average number of living offspring born to women who have completed their reproductive lifespan. In traditional societies where contraception is little if ever practised, TFRs vary enormously but rarely are they higher than the levels achieved by so called 'frontier' populations of the 19th and 20th century, such as the Amish and Hutterite sects of the United States (Campbell & Wood, 1988). Table 5.1 shows the range of TFRs for traditional societies in different regions of the world, and for frontier populations in North America.

Although social factors, such as age at marriage and the practice of 'stopping behaviour' in response to achieving desired family size are known to be important determinants of TFR in many populations (Campbell & Wood, 1988; Coleman & Salt, 1992), it is also widely accepted that bio-behavioural factors such as breastfeeding, physical activity, stress, nutrition and energy balance are important (Rosetta, 1990). There is now considerable evidence that fertility of women in traditional societies, as indicated by some measure of resumption of ovulation, is influenced by diet, exercise, and lactation (Bongaarts, 1980; Howie & McNeilly, 1982; McNeilly, Glasier & Howie, 1985; Lunn, 1985, 1988, 1992, 1994; Ellison, Peacock & Lager, 1989; Ellison, 1991, 1994; Fink et al., 1992; Panter-Brick, Lotstein & Ellison, 1993; Rosetta, 1993). Furthermore, birth seasonality has been documented in a number of rural populations including agriculturalists in present day Bangladesh (Stoeckel & Chowdhury, 1972; Becker, 1981), the Gambia (Billewicz & McGregor, 1981), Kenya (Ferguson, 1987), Lesotho (Huss-Ashmore, 1988), Tanzania (Bantje, 1988), Cameroon (Tembon, 1990), Zaire (Bailey et al., 1992), and pastoralists in Kenya (Leslie & Fry, 1989) and the Negev desert of Israel (Guptill et al., 1990). Although birth seasonality has been associated with temperature

Table 5.1. *The range of total fertility rates reported for traditional populations and frontier religious sects in North America*

	Total fertility rate	
	Lowest	Highest
East and southeast Asia	5.0	5.7
Africa	4.6	6.7
South and west Asia	3.7	7.5
Pacific (including Australia)	4.3	7.8
South America and the Caribbean	5.4	8.2
North America (Amish, Hutterites and Mormons)	6.3	9.8

Source: From Campbell & Wood (1988).

variation in Bangladesh (Becker, 1981; Becker, Chowdhury & Leridon, 1986a), this could also be associated with ecological stresses such as food availability and physical workload, which also change across the year and are a function of the agricultural cycle, which in turn is related to environmental factors such as temperature. Direct relationships between ecological factors and birth rate have been put forward for nomadic Turkana (Leslie & Fry, 1989) and agriculturalists in Lesotho (Huss-Ashmore, 1988), Zaire (Bailey *et al.*, 1992) and the Gambia (Ulijaszek, 1993b).

In rural third-world economies, work output and subsistence performance are directly related to nutritional well-being, and it is likely that the balance of energetics-related factors including nutritional state, physical workload and lactation play fundamental roles in reduced fertility, operating through reduced ovulatory function. Furthermore, women continue to be active in subsistence through much or most of pregnancy and lactation. Indeed, there are societies where women are rarely out of either state across their reproductive lifespan. In this chapter, the role of energetics is examined in: (1) the regulation of fertility; (2) the outcome of pregnancy, including birth size and survivorship; and (3) the efficiency of lactational performance. Overall reproductive performance is considered in the context of the reproductive lifespan, in which fertility regulation, pregnancy outcome and lactational performance are meshed together, with two long-term outcome variables: total fertility rate, and the women's nutritional state at the end of their reproductive careers. Evidence will be sought for maternal depletion syndrome, the postulated decline in maternal nutritional status across successive pregnancies (Cleland & Sathar, 1984), in different populations of the developing world.

The female reproductive lifespan

The female reproductive lifespan is framed by menarche, which marks its onset, and menopause, which signals its completion. Typically, the median average age at menarche in well-off, or industrialised, nations is much lower than in developing countries. Values reported for Europeans and descendants of Europeans abroad are below 14 years for all groups except Romanian village dwellers, while rural third-world populations, with the exception of Aymara living at an altitude of 3600 metres in Bolivia (13.4 years), have values above 14 years (Eveleth & Tanner, 1990), with the highest values of 20.9 and 18.8 years for the Gainj (Wood, Johnson & Campbell, 1985) and Bundi (Malcolm, 1970) of Papua New Guinea (PNG), respectively. Although an improvement in nutritional status at a population level may result in an earlier median age at menarche (Bongaarts, 1980; Tanner, 1989), and with it additional time during which females are fecund, this is unlikely to have any great effect on TFRs in developing countries, since demographic analysis across a number of populations has shown the contribution of menarcheal age to TFR to be small (Campbell and Wood, 1988). This is because social factors, such as age at marriage, play an overriding role in determining entry into reproductive life. However, an exception to this is the high level of pregnancy among adolescent girls in the United States, where inadequate weight gain in early pregnancy can significantly effect the birthweight of the infant (Scholl *et al.*, 1990). In the less-industrialised world, one can surmise that if the enforcement of virginity prior to marriage became less effective, or that marriage as an institution became less favoured, then earlier onset of menarche might have some impact on TFR, although this has yet to be shown.

The end of the female reproductive lifespan comes with the menopause. This is not a discrete event in the way that menarche is, and it is, therefore, more difficult to determine at population level. However, estimates of the average age of menopause in developing countries include 44 years for poorly nourished women in India (Wyon, Finner & Gordon, 1966) and 52 and 53 years for the general populations of Cameroon and Nigeria, respectively (Leidy, 1993). Although this is the biological end-point, terminal abstinence, the permanent sexual abstinence practised by women prior to the onset of menopause (Caldwell & Caldwell, 1981), can act as a 'cultural menopause' (Cavalli-Sforza, 1983). This practice has been postulated to be associated with the onset of grandmotherhood (Caldwell & Caldwell, 1977; Ware, 1979; Adeokun, 1982), to prevent grandmothers from competing with their own offspring in reproduction. Although it has been suggested that terminal abstinence may restrain fertility rates at older ages (Caldwell & Caldwell, 1977; Ware, 1979; Orubuloye, 1981), this is not the case in contemporary Nigeria and Cameroon (Leidy, 1993).

Between menarche and menopause, the vast majority of women in any population are able to conceive children, although the outcome in terms of total, or completed fertility rate is modulated by both biological and cultural factors. Of the cultural factors, age at first marriage is the most important, although practise of abstinence behaviours must also play a part. Of the biological factors, breastfeeding, energy expenditure, nutritional status and energy balance are known to play a part. Although it is not clear what the relative importance of each of these factors is, it is likely that the balance will vary with ecological setting, and from woman to woman.

Energetics and ovulation

The influence of energetic stress on the disruption of menstrual function has been demonstrated in a number of contexts, including athletes performing endurance exercise (Dale, Gerlach & Wilhite, 1979; Lutter & Cushman, 1982; Sanborn, Martin & Wagner, 1982; Shangold & Levine, 1982; Bullen *et al.*, 1985; Ellison & Lager, 1985; Veldhuis *et al.*, 1985; Rosetta, 1993; Cumming, Wheeler & Harper, 1994) and women experiencing weight loss (Lager & Ellison, 1990; Warren, 1990) in industrialised countries, as well as in women in developing countries (Toriola, 1988; Ellison, Peacock & Lager, 1989; Panter-Brick *et al.*, 1993). Sanborn *et al.* (1982) were the first to point out that the prevalence of menstrual dysfunction in athletes varies according to the type of sport pursued. Therefore, not all types of exercise are equivalent; Table 5.2 gives the prevalence of amenorrhoea and irregular menses in different types of female athlete. Regular runners have the highest prevalence, with cyclists and swimmers having much lower prevalences. Indeed, one group of swimmers examined had a similar prevalence of menstrual disorders to sedentary controls from another study. Although these figures are likely to be an over-estimate of real prevalences, because many studies may be biased toward greater participation by athletes with problems (Cumming *et al.*, 1994), if the degree of bias were similar among different types of athlete, then the difference between runners and swimmers and cyclists remains. From this, it would appear that physical activity as an all-encompassing variable is unlikely to be an explanatory variable, without disaggregating its components, which include intensity, duration and type of activity. Furthermore, the associated effects of energy balance, energy flux and psychological stress on menstrual function must be taken into account.

The implications of this for the study of ovulatory function of women in developing countries are that the use of female athletes as a model for the

Table 5.2. *The prevalence of menstrual dysfunction in female athletes*

Group	Oligomenorrhoea or amenorrhoea (%)	Reference
Runners	34	Dale *et al.*, 1979
Runners	26	Sanborn *et al.*, 1982
Joggers	23	Dale *et al.*, 1979
Marathon training	23	Lutter & Cushman, 1982
Elite marathon runners	19	Glass *et al.*, 1987
Marathon training	18	Shangold & Levine, 1982
Cyclists	12	Sanborn *et al.*, 1982
Swimmers	12	Sanborn *et al.*, 1982
Unselected runners	5	Sperof & Redwine, 1979
Sedentary controls	4	Dale *et al.*, 1979
Swimmers	3	Fauno *et al.*, 1991

effects of energetic stress on reproductive function may not be appropriate, beyond certain limitations. Rosetta (1992) suggested that there are great similarities between endurance athletes in the west, and third-world women, as far as ovulation is concerned, sufficient for them to be used as models for the study of the energetics of ovulation impairment in women in developing countries (Table 5.3).

The similarities between endurance athletes and women in the rural developing world include body composition, type of diet if largely vegetarian, and a possible negative energy balance for at least some of the time. Differences include breastfeeding status and the nature of menstrual cycle performance (Rosetta, 1992) but may also include differences in the type and nature of physical activity. Furthermore, the influence of psychological stress on menstrual function of women in developing countries is unknown, while among athletes, psychological depression has been shown to provide the strongest independent contribution to reproductive dysfunction after the weight/height ratio (Galle *et al.*, 1983). Rosetta (1990) put forward a model that incorporated various factors which have been shown to influence menstrual function of women in both contexts (Fig. 5.1), and this has been a useful heuristic device for attempting to understand the relationships between the various ecological factors that influence ovulation of women in developing countries.

Relationships between energy balance, stress, total daily energy expenditure and menstrual function

The Rosetta (1990) model suggests ways in which different factors may be associated in influencing reproductive function. Testing such a integrative

Table 5.3. *Similarities between endurance athletes and third-world women with high incidence of menstrual disorders*

	Endurance athletes	Third-world women from very low-income background
Breastfeeding	No	Yes
Body composition	Low weight for height Low fat percentage	Low weight for height Low fat percentage
Menstrual cycle status	High incidence of dysfunction from possible short luteal phase to secondary amenorrhoea	Long duration of post-partum amenorrhoea, lower level of steroid hormones than Western controls
Physical activities	High level of training	High level of walking, carrying loads, foraging or taking part in field work, or unskilled workers
Food intake Vegetarian	Low level of fat intake Poor quality of protein	Low level of fat intake Low level of zinc intake Poor quality of protein with lysine as usual limiting amino acid
Non-vegetarian	Possible negative energy balance	Possible negative energy balance

Source: From Rosetta (1992).

model is likely to prove extraordinarily difficult in the developing countries, but some of the relationships can be examined by using athletes in western countries as a model. This was done by Fage (1993) when she investigated the associated effects of a high level of energy expenditure on menstrual cycle length in Cambridge oarswomen. Potential effects of undernutrition were removed by choosing an elite group of athletes, the Cambridge University Women's Boat Club, in the training period leading up to the Oxford–Cambridge University Boat Race.

Twelve non-contracepting oarswomen completed questionnaires concerning their menstrual cycle characteristics, and concerning stress using the Perceived Stress Questionnaire (PSQ) (Levenstein *et al.*, 1993). The PSQ scored stress levels in the following seven categories: harrassment, overload, irritability, lack of joy, fatigue, worries and tension. These categories were pooled to give an overall stress score which was used in the analysis. Anthropometry was carried out using standard methodology

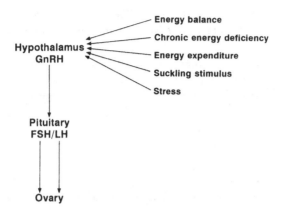

Figure 5.1. Ecological factors known to influence human ovarian function (modified from Rosetta, 1990).

(Weiner & Lourie, 1981). Measurements taken included weight, height and biceps, triceps, subscapular and suprailiac skinfolds. The percentage of body fat was calculated from the sum of the four skinfolds using the formula of Durnin and Womersley (1974) in conjunction with the Siri equation (Siri, 1956). BMI was calculated from weight and height, and FFM from weight and the percentage of body fat. Table 5.4 gives mean physical characteristics at the beginning of the study, which was close to the start of training, and 47 days later.

There was a slight reduction in body weight, as a result of a reduction in mean body fatness of 0.9 kg and an increase in mean FFM of 0.4 kg. This is similar to the sort of changes observed in women in some agricultural communities in developing countries on entering the season of highest work output. Mean menstrual cycle length was 28.4 days in the period prior to training, the Christmas vacation, a time in which these women did not engage in heavy workloads. During training, mean cycle length increased to 30.2 days for the ten women who had menstrual cycle lengths of less than 60 days. However, two of the women were anovulatory throughout the training period.

Of the variables weight, weight change, weight of fat, change in weight of fat, FFM, change in FFM and perceived stress, only weight change ($r = -0.77$, $p < 0.01$) and weight of body fat ($r = -0.68$, $p < 0.05$) were significantly associated with the length of menstrual cycle, although change in body fatness approached significance ($r = -0.61, p < 0.10$). This suggests that even in well-nourished women, small changes in body weight and body

Table 5.4. *Changes in body weight, composition and menstrual cycle length across 47 days of standardised intensive training in Cambridge women rowers*

	Time of measurement (days)	
	0	47
Body weight (kg)	68.0 (9.9)	67.5 (10.6)
Fat-free mass (kg)	50.4 (4.8)	50.8 (4.0)
Fat (kg)	17.6 (2.2)	16.7 (1.6)
Menstrual cycle length (days)	28.4 (4.8)	30.2 (4.7)[a]

Results given as means with the standard deviation in parentheses for a sample size of 12.
[a] Sample size reduced to 10 because two women became anovulatory.

fatness are associated with changes in menstrual function, while perceived stress shows no association.

In an examination of the associations between menstrual characteristics, body composition, energy expenditure and stress in 15 well-nourished Cambridge students, Small and Ulijaszek (1993) found the following. Measurements of both total daily energy expenditure and BMR gave a mean PAL value of 1.82, suggesting high levels of physical activity by the FAO/WHO/UNU (1985) criteria, with a large range (PAL minimum: 1.49; maximum: 2.63). BMI also showed great variation, ranging from 18.3 to 26.3. Menstrual cycle length was positively associated with stress ($r = 0.60$, $p < 0.01$) and negatively with body fatness ($r = -0.45$, $p < 0.05$), but not with PAL. Therefore, in the absence of nutritional stress, perceived psychological stress (measured using the PSQ) and body compositional variables are more important than energy expenditure, at least in this group of Cambridge women.

These two simple studies show ways in which the Rosetta (1990) model can be used as a framework for examining the interrelationships between different variables that can influence ovulatory function. Fage's study (1993) shows the importance of changes in body composition relative to stress in influencing menstrual function; Small and Ulijaszek (1993) show the importance of stress and body composition relative to energy expenditure in influencing the same variable. In both studies, nutritional stress was absent, and in the latter, subjects were in energy balance.

Post-partum amenorrhoea

Duration of post-partum amenorrhoea in traditional societies is much longer than in industrialised populations. For example, post-partum amenorrhoea may last 16 months for the Sundanese of West Java (Takasaka, 1986), while in Sweden and the USA it lasts 6 months on average (Soysa, 1987). In the absence of breastfeeding, menses return shortly after birth in both well-nourished and poorly nourished populations (Bongaarts, 1980), while post-partum variation in the proportion of women breastfeeding explains about 85% of between-population variance in the proportions of women menstruating (Billewicz, 1979).

Breastfeeding profoundly inhibits reproductive function (Kennedy & Visness, 1992; McNeilly, Tay & Glasier, 1994), with the intensity of suckling stimulus as the major factor which regulates the return to ovarian cyclicity (Loudon, 1987). The inhibitory effect of breastfeeding on ovulation and menstruation is the result of a neurally mediated hormonal reflex system initiated by the suckling stimulations of the nipple (Delvoye, Delogne-Desnoeck & Robyn, 1980), leading to suppression of luteinising hormone (LH) release (McNeilly, 1988) by the inhibition of the hypothalamic release of gonadotrophic releasing hormone (GnRH) (McNeilly, Howie & Glasier, 1988; McNeilly, 1994). However, the link between the suckling stimulus activating nerve terminals in the nipple and the disruption of the pattern of release of GnRH from GnRH neurons in the hypothalamus remains unknown. Although high dopaminergic tone within the hypothalamus associated with feedback effects of high plasma concentrations of prolactin have been suggsted as a possible mechanism, this does not appear to be the case (McNeilly *et al.*, 1994), since the treatment of breastfeeding women with a dopamine antagonist was shown to cause a large release of prolactin without affecting follicle stimulating hormone (FSH) or the pulsatile release of LH (Tay, Glasier & McNeilly, 1993). Opiates are known to suppress GnRH release and are associated with a decrease in GnRH pulse frequency caused by progesterone during the luteal phase of the menstrual cycle (Quigley & Yen, 1980; Ferin, van Vugt & Wardlaw, 1984). However, there is no evidence to support a role for opiates in suckling-induced suppression of GnRH release (Kremer *et al.*, 1991; Tay *et al.*, 1993).

Marked alterations in suckling times occur immediately after the introduction of supplementary food, at least when the energy density of the supplement is not low (Howie & McNeilly, 1982). In places where the nutrient value of supplements is low, the impact of supplements on overall suckling patterns may not have a major effect on the suppression of fertility (Gray *et al.*, 1990). However, the timing of the introduction of supplementary food to the breastfeeding infant cannot be discarded as a possible factor contributing to the duration of post-partum amenorrhoea.

The 'baby in the driving seat', and beyond

In 1985, Lunn introduced the concept of 'the baby in the driving seat' to explain some of the associations between breastfeeding and lactational infecundity observed in Gambian women (Lunn, 1985). Dietary supplementation of Gambian women either during lactation alone or during both pregnancy and lactation resulted in considerable shortening of the length of post-partum infecundity and reduced time to conception. These changes were associated with reductions in plasma prolactin concentration, and it was argued that the extra food had improved the milk supply to the infant, who in turn suckled less vigorously. The less intense suckling had in turn decreased ovarian suppression, and it was suggested that, in developing countries where maternal nutritional status was poor, a hungry baby could, by suckling more frequently or for longer, stimulate a greater release of prolactin that drove maternal metabolism towards milk production and strengthened the inhibitory effect on ovarian activity (Lunn *et al.*, 1984; Lunn, 1985).

In a recent re-evaluation, Lunn (1994) modified his view in relation to the central role of prolactin in the scheme of suckling-induced ovulatory inhibition. This brought the 'baby in the driving seat' hypothesis in line with the Rosetta (1990) model. The revision is based on a reinterpretation of energetics data collected in Gambian women in the light of newer methods of measuring energy expenditure and changes in the theoretical understanding of energy balance in lactation (Chapter 3). Reported energy intakes of lactating Gambian women have been well below FAO/WHO/ UNU (1985) recommendations but appeared to be compatible with successful lactation, which led to speculations about the existence of energy-saving adaptations which could help the maternal metabolism accommodate the cost of lactation (Prentice, 1980). However, little evidence for such adaptation has been found. Gambian women do not have low, or down-regulated BMR in comparison with British women in Cambridge (Coward, Goldberg & Prentice, 1992), and lactating Gambian women exhibit only a small reduction in BMR compared with non-pregnant, non-lactating Gambian women (Lawrence & Whitehead, 1988). Furthermore, their energy expenditure, measured by the doubly labelled water method, was about twice the level that had been reported for intake in earlier studies (Singh *et al.*, 1989), while their PAL was 1.92, suggesting a heavy workload (FAO/WHO/UNU, 1985). Although some of the difference could result from a real negative energy balance, since the comparison between intake and expenditure was carried out in the wet season (Singh *et al.*, 1989) when weight loss traditionally occurs (Ulijaszek, 1993b), the discrepancy can only account for at most half of any of the monthly weight

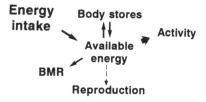

Figure 5.2. Energy partitioning in training athletes (from Lunn, 1994).

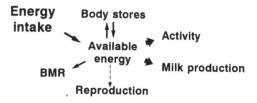

Figure 5.3. Energy partititioning in lactating women in developing countries (from Lunn, 1994).

loss recorded in the study in the 1980s, as documented by Cole (1993). Lunn (1994) concludes that estimates of maternal energy intake in the Gambia, and probably those from other countries too, must be incorrect, as they are incompatible with energy utilisation. Supporting this view is evidence that supplementing the maternal diet with 3 MJ dietary energy per day had minimal effects on body weight or composition, but resulted in increased fecundity (Lunn *et al.*, 1984). Apart from relieving the energy stress of lactation, the extra energy is surmised to have been used in increased work activity (Lunn, 1994). Therefore, lactating Gambian women are probably marginally energy deficient, work hard but show little metabolic adaptation to the energy needs of lactation.

Comparing the energy partitioning of training athletes and lactating women in the developing world (Figs. 5.2 and 5.3), Lunn (1994) argues that the infant-stimulated drive to milk production could take precedence over ovarian function, especially in the rural third world, where there is high energy expenditure in physical activity. Furthermore, he suggests that the nutritional stress of milk production could lead to lactational amenorrhoea irrespective of the suckling reflex arc. When the nutritional drive to milk production is reduced or ceases, nutritional stress is also eased, and reproductive function can quickly return; thus the role of prolactin is marginalised relative to the overriding maternal energetic stress. (Fig. 5.4) (Lunn, 1994). However, recent work by Palmon *et al.* (1994) has shown that

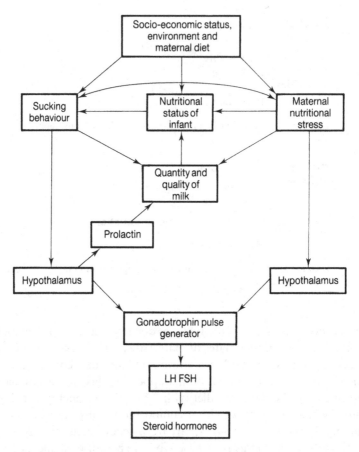

Figure 5.4. Interrelationships between energetics and fecundity in lactating women (from Lunn, 1994).

the gene for GnRH is expressed in the mammary glands of lactating rats.

Three factors do not appear on the diagram and are worthy of inclusion. The first is the importance of maternal activity level and its impact on maternal nutritional status; high activity level also means high energy flux, and this could be operating independently on the hypothalamic GnRH pulse generator. The second is energy balance. Gambian women exhibit seasonal weight change (Cole, 1993) as a consequence of negative energy balance in the wet season (Ulijaszek, 1993b); studies in Zaire (Ellison *et al.*, 1989) and Nepal (Panter-Brick *et al.*, 1993) have shown weight loss to be associated with reduced ovarian function. In the Gambia, Billewicz and

McGregor (1981) demonstrated seasonality of births, while Ulijaszek (1993b) suggested that this birth seasonality could be attributed to weight loss caused more by low energy intakes (estimated by difference between energy expenditure and balance, not intake) than higher expenditure during the wet season. This does not diminish the potential importance of high work output in these women, since total energy expenditure is high across all seasons but is about 1 MJ/day higher in the wet season than in the dry (Lawrence & Whitehead, 1988). The third factor not included in Fig. 5.4 is psychological stress; the relationships between this variable and menstrual function has been little examined in western women, and not at all in women in developing countries. The Rosetta model incorporates stress as a factor influencing reproductive function.

There is a need to describe the variation in ovulatory function of women in developing countries and to determine the relative importance of different factors in creating this variation. Although a great deal of variation at the population level may result from breastfeeding, the Bongaarts (1982) model does not incorporate measures of energetics or of body size or composition. Therefore, an unknown proportion of the variance in post-partum amenorrhoea ascribed to breastfeeding could be caused by other factors not considered by demographers. Future studies should examine the possible relationships between ovulatory function, lactation, maternal energy balance and energy expenditure and the timing of supplementary feeding to the breastfeeding child. Furthermore, there is a need to develop culturally appropriate indices of psychological stress for use in non-European contexts. The use of biochemical markers of stress might also be useful. Plasma and urinary adrenaline can reasonably measure levels of arousal, be they caused by anxiety, alertness, excitement or distress (Aslan *et al.*, 1981; Akerstedt *et al.*, 1983; Pollack & Steklis, 1986; O'Donnell *et al.*, 1987; Fibiger & Singer, 1989; Oleshansky & Meyerhoff, 1992). However, the use of urinary cortisol for stress measurement is less certain (Kirschbaum & Hellhammer, 1989), although the balance of evidence suggests that negatively connotated arousal can cause increased cortisol secretion (Herbert *et al.*, 1986; Vickers, 1988; Hubert & de Jong-Meyer, 1989).

Pregnancy and pregnancy outcome
In well-nourished populations, the energy cost of pregnancy has been estimated to be about 335 MJ (Hytten, 1980). This is a value that the FAO/WHO/UNU (1985) has adopted for international use, although 75% of this value is deemed by them to be reasonable for the energy requirement of pregnancy for healthy women who reduce their physical activity

Table 5.5. *Components of maternal weight gain in pregnancy, for Hytten's 'standard' pregnant woman*

	Weeks of pregnancy				
	10	20	30	40	Total
Protein					
Fetus and placenta	2	43	220	540	
Mother	33	122	278	385	925
Fat					
Fetus and placenta	–	3	83	444	
Mother	328	2060	3510	3380	3824

Source: From Hytten (1980).

(FAO/WHO/UNU, 1985). Total weight gain across pregnancy of Hytten's 'standard' pregnant woman is 12.5 kg, the components of which are shown in Table 5.5.

It should be noted that the 'standard' pregnant woman is an idealised construct, analogous to the 'reference' male of past generations, who weighed 75 kg in the United States and 55 kg in India, and the '50th centile' child of the present day. The components of weight gain pregnancy cannot possibly be as precise as is suggested by Table 5.5, since measurements of most of them cannot be made to this degree of precision. However, they indicate that the fetus gains protein and fat in more-or-less equal proportion, while the mother gains about an equal proportion of protein to the fetus, but more than seven times more fat. This estimate of body fat increase is subject to considerable uncertainty, given that existing body-composition techniques, save for assays of total red cell mass, measure the entire feto-maternal unit and not its components (Forbes, 1987).

This description of 'normal' pregnancy is refuted by a growing body of data on the energy cost of pregnancy, estimated in both developing and industrialised nations and using a range of increasingly accurate methods. There is evidence to suggest that the energy cost of pregnancy is considerably less than the value of 355 MJ, across pregnancy. Table 5.6 gives estimates of the energy cost of pregnancy in five countries (Durnin, 1988), and shows it to be 23% (Gambia), 54% (Philippines), 62% (Thailand), 76% (Netherlands) and 84% (Scotland) of the Hytten (1980) estimate. Lower than expected BMRs form a significant component of these reduced estimates, at least in the Gambia, Thailand and the Philippines (Durnin, 1988).

Table 5.6. *Energy cost of pregnancy in different populations*

Country	Energy cost (MJ/pregnancy) of:					
	Fetus	Placenta	Maternal fat	BMR	Everything else[a]	Total
Scotland	33.9	3.1	105.6	125.9	12.1	280.9
Netherlands	34.4	3.1	59.8	144.3	12.3	254.1
Thailand[b]	29.9	2.5	64.6	100.4	10.4	207.6
Philippines[c]	28.9	2.5	59.8	79.5	10.1	180.8
Gambia	29.9	2.3	27.6	7.9	10.4	77.7

Source: From Durnin (1988).
[a] Including uterus, breasts, blood volume, amniotic fluid and extracellular fluid.
[b] Estimated from values at 10 weeks of gestation.
[c] Estimated from values at 13 weeks of gestation.

The nutritional state of the mother may influence BMR during pregnancy. Gambian women receiving a dietary supplement of about 1.9 MJ/day showed a total increment in BMR above non-pregnant values of 54 MJ across the entire pregnancy, in comparison with non-supplemented controls who showed an increment of only 5 MJ (Lawrence *et al.*, 1984). Gambian women in the first trimester of pregnancy had a slightly lower sleeping metabolic rate per kg FFM than did non-pregnant, non-lactating women and women at later stages of pregnancy (Heini *et al.*, 1992). In addition, BMR of a longitudinal sample of pregnant Gambian women showed an overall decline below non-pregnant values of about 5%, returning to pre-pregnancy levels by 18 weeks of gestation (Poppitt *et al.*, 1993). There is no equivalent reduction in BMR in well-off populations in Scotland (Durnin, 1987), the Netherlands (van Raaij *et al.*, 1990) or England (Goldberg *et al.*, 1993).

The energy cost of performing standardised tasks declines across pregnancy in the Dutch, Scottish, Thai and Gambian study groups, values being lower in Gambian and Thai women than in Dutch and Scottish women at similar stages of pregnancy (Durnin, 1988). However, the energy cost of common daily activities in Gambian women does not increase upon receiving dietary supplementation (Lawrence *et al.*, 1985), suggesting that these effects may not be caused by poor nutritional status, at least in Gambian women.

Table 5.7 shows total weight gain across pregnancy for a number of populations, both industrialised and non-industrialised. This table includes the Thomson and Billewicz (1957) data on British primigravidae

Table 5.7. *Total weight gain across pregnancy: various studies*

Country	Total weight gain (kg)	Period covered	Reference
Britain	12.5		Thomson & Billewicz, 1957
England	11.9	Pre-pregnant–36 weeks	Goldberg *et al.*, 1993
Scotland	11.7	10 weeks–term	Durnin, 1987
Netherlands	10.5	10 weeks–term	Durnin, 1987
Thailand	8.9	10 weeks–term	Durnin, 1987
Philippines	8.5	10 weeks–term	Tuazon *et al.*, 1987
Philippines	8.4	Pre-pregnant–term	Siega-Riz & Adair, 1993
Taiwan	7.6		Adair, Pollitt & Mueller, 1983
Gambia	7.4	10 weeks–term	Lawrence *et al.*, 1987
India	7.0		Tripathi *et al.*, 1987
Guatemala	7.0		Lechtig *et al.*, 1975
India (Gujarat)	6.1		Anderson, 1989
East Java	6.0	1–3 and 7–9 months	Hull, 1983
India (Maharashtra)	5.7		Anderson, 1989
Bangladesh	4.8		Krasovec, 1989

Source: Taken from Siega-Riz & Adair (1993).

who, eating to appetite, gained 12.5 kg, and upon which the Hytten (1980) recommendation is based. Mean values for all other groups reported are below 12.5 kg, with the three European populations being closest to this value. English, Scottish and Dutch women show mean weight gains across pregnancy of 11.9, 11.7, and 10.5 kg, respectively. Mean weight gains in pregnancy for the various groups of women in developing countries are lower than this, with Gambian, Taiwanese, Guatemalan, Indian and East Javanese women showing weight gains of less than 8 kg.

Although some component of this lower weight gain might carry no additional risk for pregnancy outcome, extremely low gains are likely to contribute to low birthweight. For the non-industrialised countries for which acceptable data are available, the prevalence of low birthweight (term infants born weighing less than 2.5 kg) varies from a high of 27% in Indonesia to a low of 9% in Nigeria (Offringa & Boersma, 1987). There is some overlap with industrialised nations, since the prevalence of low birthweight ranges between 9% for the USA to just over 2% for (South) Korea. Fetal growth retardation probably contributes more than prematurity (gestation length 37 weeks or less) to elevated neonatal mortality in developing countries (Haas, Balcazar & Caulfield, 1987), with no fewer than 70% of low birthweight infants being full-term (gestation length between 38 and 42 weeks) and growth retarded (Villar & Belizan, 1982a).

It is possible to distinguish between full-term growth-retarded infants according to whether they have experienced chronic or acute growth failure *in utero* (Villar & Belizan, 1982b). Proportionately growth-retarded are those that experienced chronic growth failure; they are both light and short, reflecting a short-fall in the accumulation of both labile tissues, such as muscle, viscera and fat, and more conservative skeletal tissues. Such infants have weights for length which are within the range of the general neonate population. Disproportionately growth-retarded infants experience acute growth failure *in utero* and show a short-fall in the development of labile tissue without a short-fall in the development of skeletal tissue. Thus, they have low weight for length (Villar & Belizan, 1982b).

Proportional growth retardation usually occurs across much or most of pregnancy and, in industrialised nations, occurs in the face of adequate energy and nutritional supply. It is commonly the result of genetic factors (Chiswick, 1985): of all proportionally retarded fetuses recorded in Britain, 5–15% have malformations, while 2% have chromosomal abnormalities (Chen & Falek, 1974; Anderson, 1976). It is debatable to what extent proportional growth retardation in developing countries is caused by similar factors. However, a number of authors (Davies *et al.*, 1979; Villar *et al.*, 1982, 1984) have suggested that proportionate growth retardation involves long-term problems of growth and behavioural development, and genetic factors should not be discounted.

Disportionate growth retardation implies a less severe or later onset restriction of growth (Chiswick, 1985) and is most likely to occur when energetic and nutrient restriction has taken place at the time of peak weight gain *in utero*, which is the third trimester of pregnancy, but not earlier. In the developing world, both proportional and disproportional intrauterine growth retardation are believed to occur in fairly similar proportions (Haas *et al.*, 1987).

In a study of early neonatal mortality rates in Mexico and Bolivia, Haas *et al.* (1987) showed great differences in death rate in the first 2 days after birth, according to newborn classification (Table 5.8). In both places, death rates were more than an order of magnitude higher among pre-term infants than among those who were of appropriate size for their gestational age. However, of greater interest are the term infants who are light for their gestational age, since they represent the children who are likely to have experienced either chronic or acute energetic/nutritional deficits when *in utero*. In both Mexico City and Santa Cruz, Bolivia, being light for gestational age carried a greater than two-fold greater risk of death than being of appropriate weight for gestational age. Discriminating between

Table 5.8. *Early neonatal mortality rates for different classifications of Mexican and Bolivian infants*

Newborn classification	Mexico City		Santa Cruz, Bolivia	
	Deaths[a]	Risk ratio[b]	Deaths[a]	Risk ratio[b]
Term	4.0		3.3	
Appropriate for gestational age	2.9	1	2.3	1
Light for gestational age	6.1	2.1	8.4	3.7
Proportional growth retardation	4.1	1.4	4.7	2.0
Disproportional growth retardation	8.3	2.9	13.2	5.7
Pre-term	54.0	18.6	117.0	50.9
Total mortality rate	9.1		12.3	
Total live births	10024		12786	

Source: From Haas *et al.* (1987).
[a] Deaths per 1000.
[b] Mortality rate for the specific newborn group relative to the rate for full-term, appropriate-for-gestational age infants.

proportional and disproportional growth retardation shows the latter to have by far the greater risk of death. Neonates having experienced acute energetic and/or nutritional stress are at two to three times greater risk of death than are those that experienced proportionate growth retardation, whether it be caused by chronic nutritional or energetic stress or by undefinable genetic factors.

Relating this back to pregnancy weight gain, it is not possible to say whether the low pregnancy weight gains in women in developing countries shown in Table 5.7 reflect patterns which result in minimal pregnancy wastage or neonatal death. Data collected from women in rural Gambia may help give some idea of the extent to which this is the case. Dietary supplementation of pregnant women was carried out in the Gambia in the early 1980s, with the aim of increasing birthweight (Prentice *et al.*, 1983). This had the desired effect in women receiving supplements during the third trimester of pregnancy, when it coincided with the wet season. However, supplementation had no impact on birthweight at other times of the year, nor if given earlier in pregnancy (Prentice *et al.*, 1987b). This was taken to mean that the energetic stress of the third trimester of pregnancy was sufficiently severe in the wet season that supplementation had an impact on birthweight (Prentice *et al.*, 1987b). Combining this with the knowledge that disproportionate growth retardation is the result of acute stress and that such disproportionate growth is associated with high neonatal mortality risk (Haas *et al.*, 1987) (Table 5.8), the Gambian supplementation trial data suggest that: (1) intrauterine growth was acutely restricted in women in the third trimester of pregnancy during the wet season; (2) this restriction was likely to have carried a high neonatal mortality risk; and (3) the energetic constraint on intrauterine growth was removed by supplementation. Therefore, the average weight gain across pregnancy of 7.4 kg in pregnant Gambian women is unlikely to be appropriate for the maximal survival of offspring, even though there is some evidence of increased metabolic efficiency compared with European populations (Durnin, 1988; Heini *et al.*, 1992; Poppitt *et al.*, 1993). Weight gain across pregnancy may also be too low for maximal neonatal survival in other groups of women, notably among those whose weight gain is considerably lower than that of European women (Table 5.7).

An obvious way of saving energy is by reducing the total amount of physical activity performed during pregnancy. In the Gambia, women become progressively less active during pregnancy (Roberts *et al.*, 1982), the main changes in activity being reductions in the amount of walking within the village and in the duration of household work, and a tendency to go less often to the fields (Lawrence & Whitehead, 1988). However, the effect of supplementation during the wet season on increasing birthweight indicates that savings in energy expended in physical activity can only partially compensate for the energetic stress experienced. Furthermore, savings obtained through the lowered energy cost of physical activity in pregnancy (Durnin, 1988) are lost again, since a greater body mass has to be moved when performing work during pregnancy.

A reanalysis of the Gambian supplementation trials using seasonally adjusted weight gains was carried out by Lawrence *et al.* (1987). Factorial analysis of the type carried out by Hytten and Leitch (1971) showed that, on average, supplementation had no effect on fetal weight (Table 5.9) but substantially increased maternal fat stores. Indeed, the analysis suggests that under non-supplemented conditions Gambian women lose about 0.6 kg of body fat during pregnancy, and only with supplementation do fat stores increase. This analysis averaged out seasonal effects, which were considerable. For example, unsupplemented non-pregnant, non-lactating women showed a 5 kg fluctuation in body weight across a year, but only a 2 kg fluctuation when supplemented (Lawrence *et al.*, 1987). Furthermore, as already stated, supplementation only increased birthweight when it took place in the wet season. Therefore, it can be surmised that there are two effects of supplementation: (1) increasing the birthweight of infants born to women experiencing acute nutritional stress during the wet season, at the

Table 5.9. *Weight gain in maternal adipose tissue stores of Gambian women, calculated by the factorial method of Hytten and Leitch (1971)*

	Weight gain (kg) when diet is	
	Unsupplemented	Supplemented
Total weight gain	7.4	9.2
Fetus	3.0	3.0
Placenta	0.5	0.5
Uterus, breasts, blood	4.5	4.0
Products of conception	8.0	7.5
Maternal fat (total minus products of conception)	−0.6	1.7

Source: From Lawrence *et al.* (1987).

stage in their pregnancy when the energetic needs of the fetus for growth are at their highest; and (2) a more generalised repletion of maternal fat stores. Fat gain during pregnancy is important in relation to lactational performance. The prospect that Gambian women may lose fat in pregnancy brings the issue of maternal depletion into focus; evidence for maternal-depletion syndrome in Gambian and other groups of women will be examined subsequently.

Lactation

The energy cost of lactation is the energy content of the milk secreted and the energy required to produce it. In order to maintain energy balance across the reproductive lifespan, child-bearing women need to have acquired body fat in pregnancy order to sustain lactation.

Table 5.10 gives estimates of body fat gain during pregnancy from a number of studies. Hytten and Leitch (1971) determined that maternal fat adipose tissue stores in European women increase by about 3.4 kg during pregnancy, the energy equivalent of this being lower than the 150 MJ of fat reserves the present FAO/WHO/UNU (1985) references assume that the average women will accumulate during pregnancy. This would contribute some 500 kJ/day toward the extra 2100 kJ/day that the FAO/WHO/UNO (1985) estimates is needed for successful lactation. In the Gambia, fat gain during pregnancy is on average only 0.6 kg (Durnin, 1988), and the role of maternal body fat as an energy source for lactation does not hold (Lunn, 1994).

Lactational performance of women in developing countries is adequate in comparison with western women (Orr-Ewing, Heywood & Coward, 1986; Prentice *et al.*, 1986a); intakes of breastmilk of infants below the age

Table 5.10. *Fat gain during pregnancy in various studies*

Country	Fat gain (kg)	Reference
England	3.4	Hytten & Leitch, 1971
Scotland	2.3	Durnin, 1988
Thailand	1.4	Durnin, 1988
Netherlands	1.3	Durnin, 1988
Philippines	1.3	Durnin, 1988
Gambia	0.6	Durnin, 1988

of 6 months follow similar patterns in Papua New Guinea, the Gambia and the UK. This raises the question of possible energy-saving adaptations that might help the maternal metabolism to accommodate the cost of lactation, where fat gains across pregnancy are low (Prentice & Prentice, 1990; Coward *et al.*, 1992). Lunn (1994) has summarised the possible sources of energy for lactation: (1) increased food consumption; (2) reduced physical activity; (3) reduced BMR; (4) reduced TEF; and (5) use of body fat stores.

In a study of lactating women in Cambridge, England, it was estimated that 62% of the energy for milk production comes from increased energy intake, 35% from decreased physical activity, and 3% from reduced BMR (Goldberg *et al.*, 1991; Lunn, 1994). Similar results have been reported for Dutch women (van Raaij *et al.*, 1990, 1991). Furthermore, Illingworth *et al.* (1986) found lower TEF in lactating women compared with bottle-feeding controls and suggested that enhanced metabolic efficiency occurred in the former group. However, the balance of energy saving for sustained lactation may vary according to ecological setting. In industrialised nations, there may be little problem in increasing energy intake, while the potential for reduced physical activity may be limited. In rural third-world women, the reverse may be true; there may be little potential to increase food intake, but greater potential to reduce energy expenditure in physical activity (Lunn, 1994).

In addition to increased energy intake and reduced physical activity, BMR and TEF, there is the possibility of utilisation of body fat stores, a simple proxy for this being weight loss across lactation. In North American women, this lactational weight loss amounts to 1.2 kg in the first 3 months of lactation, 3.4 kg in the first 6 months, and 4.4 kg across 12 months of lactation (Dewey, Heinig & Nommsen, 1993). In the Philippines, the average maternal weight loss over the first 6 months of lactation is 1.5 kg (Guillermo-Tuazon *et al.*, 1992), while in Papua New Guinea the loss is 1.6 kg over 12 months of lactation (Orr-Ewing, 1985); in both places, the utilisation of fat stores in support of the energy cost of lactation is lower

Table 5.11. *Probable maximum energy savings in lactating women in developing countries*

	Reduction (kJ/day)
Physical activity (workload)	500–1500
BMR	500
TEF	200
Total saving	1200–2200

Source: From Lunn (1994).
Energy content of 750 ml breastmilk is 2100 kJ.

than among women from industrialised countries. Where the weight gain in pregnancy is low, such as in the Philippines, fat utilisation for lactation is also likely to be low. Indeed, Prentice and Prentice (1990) have argued the extreme case that the concept that fat stores are automatically used to supplement lactation may not be valid in developing countries. If intake does not increase, then the balance must come from any or all of the methods of reducing energy expenditure: physical activity, BMR, and/or TEF. Table 5.11 gives an estimate of the probable maximum energy savings which could be made to support lactation. The energy content of 750 ml of breastmilk (about the average daily amount that women produce in the first 3 months of lactation in both industrialised and developing countries) is about 2100 kJ and could just about be achieved in the absence of increased food intake, predominantly by reduced levels of physical activity.

At lower levels of nutritional adequacy, women may reduce their levels of physical activity to conserve energy. Energy savings of about 15% resulting from reduced activity have been reported in pregnant and lactating women in India (McNeill & Payne, 1985), Nepal (Panter-Brick, 1989) and the Gambia (Lawrence & Whitehead, 1988). However, such a reduction would have to be prolonged in order to sustain the energy cost of lactation without overall negative energy balance upon completion of lactation. In the Gambia, maternal workload was only markedly diminished during the first few weeks of lactation; after 3 months no decrease was detected (Lawrence & Whitehead, 1988). By this time, women had resumed working long hours in the fields at high levels of energy expenditure, as determined by PAL (Singh *et al.*, 1989). In rural Gambia, women cannot be absented from work for any period of time, since the nutritional well-being of their family depends on them working when there is work to be done. The potential for metabolic adaptation in lactating Gambian women is limited, reduction in BMR being able to account, at most, for a quarter of the energy content of an adequate breastmilk supply. Furthermore, TEF has been sought for in this group of women and not found (Frigerio *et al.*, 1992). Thus, if the energy cost of lactation cannot be sustained without some energetic imbalance, this raises the possibility of maternal depletion of bodily tissue for successful lactation.

Under conditions of moderate energy restriction, it has been suggested that lactational performance of Gambian women does not suffer (Roberts *et al.*, 1982), and although breastmilk output has been shown to decline with duration of breastfeeding, it is not clear to what extent this is confounded by the effects of supplementary or complementary feeding of Gambian infants. Severely malnourished lactating women may show one of two patterns of breastmilk production: (1) an inability to increase initial milk production to keep up with the demand of the child; and (2) experiencing 2 to 3 months of abundant lactation, followed by a fall in output (Chavez & Martinez, 1980).

The level of breastmilk production in any woman is likely to be the result of complex interactions between the size and suckling abilities of the newborn baby, maternal nutritional status, the growth performance of the breastfeeding child and possible genetic factors. The amount of breastmilk produced by mothers has been found to be related to the size of the child being breastfed (Whitehead *et al.*, 1978; Orr-Ewing *et al.*, 1986). Prentice *et al.* (1986a) showed that Gambian mothers have characteristic levels of milk output in successive lactations, and that such variation is maintained in the lactational performance of daughters when they reach reproductive age.

Maternal-depletion syndrome

Maternal depletion is not a universal phenomenon in developing countries, although various effects of increasing parity on pregnancy outcome and lactational performance have been documented. However, in examining evidence for such long-term energetic stress, one must be cautious not to attribute the effects of natural ageing to maternal depletion. Since the process of natural ageing is poorly characterised in developing countries, this makes the study of maternal depletion difficult.

Lactational performance of Gambian women declines with parity and age, although there is no change in body weight or BMI (Table 5.12). The energy density of their breastmilk falls by about 8% between parities 1 and 9 +. Protein and fat concentrations fall by 11% and 21%, respectively, across the same range of parities, while lactose concentration rises by 4%. However, there is little effect of parity on breastmilk volume (Ann Prentice, 1986). This variation in lactational performance is reflected in the growth of Gambian infants aged 1 to 3 months.

Table 5.12. *Parity and lactational performance in Gambian mothers*

Parity	Age (years)	BMI (kg/m²)	Breastmilk composition			
			Energy (kJ/100 g)	Protein (g/100 g)	Fat (g/100 g)	Lactose (g/100 g)
1	22.1	21.8	295	1.11	4.54	6.66
2	23.9	20.5	287	1.11	4.07	6.74
3 to 8	32.1	21.1	277	1.03	3.71	6.75
9 or more	38.5	22.7	273	0.99	3.59	6.93

Source: From Ann Prentice (1986); Prentice *et al.* (1981); Prentice *et al.* (1980).

Firstborn babies have significantly lower birthweights than infants of parities 2 to 8, but have caught up with them by the age of 3 months. Children born to mothers of very high parity have birthweights which are slightly lower than those of parities 2 to 8, but their early growth is poor and by 3 months of age they have significantly lower weights than their lower parity counterparts (Ann Prentice, Cole & Whitehead, 1987a). This variation in lactational performance and infant growth with parity is of the same magnitude as that displayed between the wet and dry seasons (Rowland *et al.*, 1981). Collectively, the data suggest that maternal depletion is avoided by reduced lactational performance with increasing parity. Although a mother may have to be extremely undernourished before her milk yield is likely to be seriously affected (Whitehead & Paul, 1988), small adjustments in lactation performance and physical activity may be the main mechanisms through which maternal nutritional status is maintained across repeated pregnancies. However, this must remain tentative until an analysis in which the confounding effects of parity and increasing age on lactational performance is carried out.

Summary
Reproductive performance has several components, including the onset of ovulatory function, fecundity, pregnancy and birthweight, and lactational performance. Body composition, energy expenditure and energy balance are important factors in the regulation of fertility in non-contracepting populations. In particular, associations between breastfeeding and lactational infecundability may be less a function of prolactin release than the energy nutritional status of the lactating mother.

Pregnancy outcome includes birth size and survivorship. Although the recommended total energy requirement for tissue gain and increased metabolism across pregnancy is currently 355 MJ (FAO/WHO/UNU,

1985), the energy cost of pregnancy is considerably less than this in women in a number of developing countries, who have lower mean weight gains in pregnancy than women industrialised nations. Although to some extent lower weight gain might carry no additional risk for pregnancy outcome, extremely low gains are likely to be major contributors to the high prevalence of low birthweight in the developing world. Dietary supplementation trials carried out with Gambian women suggest that intrauterine growth is acutely restricted in women in the third trimester of pregnancy during the wet season, and that this is likely to carry a high neonatal mortality risk. Furthermore, the energetic constraint on intrauterine growth is removed by supplementation. Therefore, the low average weight gain across pregnancy of Gambian women is unlikely to be appropriate for the maximal survival of offspring, and this may also be true for women in other non-industrialised societies.

The energy cost of lactation is the energy content of the milk secreted and the energy required to produce it. In order to maintain energy balance across the reproductive lifespan, child-bearing women need to have acquired body fat in pregnancy in order to sustain lactation. Lactational performance of women in developing countries appears to be adequate in comparison with women in industrialised nations. The possibility that energy-saving adaptations might help the maternal metabolism to accommodate the cost of lactation under conditions where fat gains across pregnancy are low has been considered by various authors. In industrialised nations, the majority of the extra energy required for breastmilk production comes from increased energy intake, with an additional major contribution from reduced physical activity. In this setting, there may be little problem in increasing energy intake, while the potential for reduced physical activity may be limited. In rural third-world women, the reverse may be true; there may be little potential to increase food intake, but greater potential to reduce energy expenditure in physical activity. The energy content of 750 ml of breastmilk (about the average daily amount that women produce in the first 3 months of lactation, in both industrialised and developing countries) is about 2.1 MJ, and, in the absence of increased food intake, could just about be achieved predominantly by reduced levels of physical activity. However, reductions would have to be prolonged in order to sustain the energy cost of lactation without overall negative energy balance on completion of lactation. This raises the possibility of depletion of maternal bodily tissue for successful lactation. However, various sources of energetics data from the Gambia suggest that maternal depletion is avoided by reduced lactational performance with increasing parity.

6   *Growth and body size*

There is great between-population variation in the rate of physical growth and development and attained body size at any given age (Eveleth & Tanner, 1990), largest differences being between populations in industrialised and non-industrialised nations, and between well-off and poorer groups within countries (Ulijaszek, 1994). This variation has both genetic and environmental components. If it can be accepted that humans have a common point of origin in Africa, with dispersion only coming late in evolutionary time, human populations, regardless of where they live, have tremendous genetic similarity. Following from this, the pattern of dispersion out of Africa suggests that variation in any trait is likely to be a continuum across populations, and that any classification of contemporary populations is to some extent arbitrary. Furthermore, prior to the onset of agriculture, some 12 000 years ago, humans lived as hunter–gatherers at low population densities often in isolation from each other, and this leads to the possibility of the development of regional population genetic differences.

Migrations taking place after the onset of agriculture served to create larger, more genetically homogeneous populations across wide areas, with genetically isolated populations left in less-hospitable ecological niches. Genetically isolated populations to be found in the world today include tribal groups of hunter–gatherers in Africa, Latin America and Asia. Large-scale colonisation of the Pacific Islands took place at this time. Later migrations, during colonial times, include the migrations of (1) Europeans to the Americas, Australasia and parts of Africa; (2) Africans, mostly of Bantu origin, to the Americas and the Caribbean; and (3) Asiatics, Chinese in particular, and Indo-Mediterraneans, largely South Asians, to most parts of the tropical world and to parts of the New World. Migrations in the post-colonial period are largely related to economics and urbanisation. Examples include the migrations of Mexicans to the USA, of South Asians to Britain, Algerians to France and Chinese to most industrialised nations. In addition, there is the global trend of rural to urban migration in post-colonial times. Thus, human populations have great homogeneity for most genetically determined characteristics; between-population variation is likely to have taken place in the recent evolutionary past. Migrations and population expansions have created several larger population groupings within which genetically isolated groups may be found.

Possible differences in growth patterns which might be attributed to genetic factors can be examined in a number of ways, including: (1) comparing measures of body size, such as height, of populations from industrialised countries and from the highest socio-economic groups in developing countries, at similar chronological ages; and (2) seeking evidence of positive secular trends in body size in different populations, and their cessation.

In examining these two bodies of evidence, Ulijaszek (1994) concluded that the growth patterns of all major population groupings are likely to have similar genetic potential, with the exception of Asiatics. However, there are no data on either secular trends or well-off groups from populations that have until recently been genetically isolated; it is not known whether they share the same potential for growth as the major populations that surround them. Furthermore, little is known about the genetic potential for growth of Aboriginal populations in Australia, or in Pacific Islands populations.

Although genetic potential for growth may be similar in many populations, the great differences in growth patterns between well-off and less well-off populations suggest that environmental factors play a significant role in shaping human growth. In particular, within- and between-population differences in physical growth in non-industrialised populations can largely be attributed to differences in nutritional well-being and exposure to infectious disease (Nabarro *et al.*, 1988; Waterlow, 1988; van Lerberghe, 1990; Neumann & Harrison, 1994). In anthropological terms, this can be considered as adjustment and adaptation to both the nutritional and disease environments; smaller body size may confer an advantage if it adjusts the size of individuals to available nutritional and energetic resources, but it may be disadvantageous in other respects, such as greater susceptibility to infectious disease, or lower physical work capacity. In this chapter, the importance of energetics to the growth process and to bodily maintenance is considered, with reference to non-industrialised populations. In addition, the functional implications of achieving smaller body size than the genetic potential are examined for both children and adults.

Becoming small

The process of becoming small has been described countless times. Usually, either attained size in height and/or weight at a given age, or growth velocity between two ages, is compared with some reference growth pattern. Currently, the NCHS (1977) references developed in the USA are used internationally. However, there is limited consensus over the use of such references. Any use of growth references internationally should acknowledge that they can act, at best, as imperfect yardsticks, since human populations may show similar growth characteristics but are unlikely to ever become so homogeneous that they show the same genetic potential for growth. Since the NCHS growth references do not represent the greatest possible human potential for growth (Ulijaszek, 1994), they may not be any more appropriate for international use than growth references developed in other developed countries. The NCHS references are poorly modelled (Cole, 1989) and are based on two different data sets, overlapping from 2 to 3 years, with a constant difference over that period between the measurements of length and height. There is a need for the data used in the development of the NCHS growth references to be reanalysed in a more sophisticated manner if they are to be of use internationally. Although new references which take these limitations into consideration may become available from the USA in coming years, it is questionable whether these are likely to be any better than existing growth references based on measurements of well-off children in the Netherlands (Roede & van Wieringen, 1985).

An international growth reference could be used for European and European-origin populations, as well as African and African-origin and Indo-Mediterranean populations. Current evidence suggests that they may not apply to Asiatic populations, but in the absence of definitive evidence of a cessation of the secular trend in any well-off Asiatic population, this assumption must remain tentative (Ulijaszek, 1994). It is not clear whether genetically isolated populations in various parts of the world including Africa, India, Latin America and Asia are likely to show the same potential for growth when placed in favourable environments. In addition, almost nothing is known about genetic potential for growth of Aboriginal populations in Australia, or in Pacific Islands populations. Despite this, the process of becoming small has to be considered in relation to some reference growth pattern, acknowledging that for some populations, the description may be inaccurate to some, usually unknown, degree. Figures 6.1–6.4 illustrate the process of becoming small for three populations, the Bundi of Papua New Guinea (Malcolm, 1970), and groups in Somalia (Gallo & Mestriner, 1980) and Zaire (van Lerberghe, 1990). Values for weight and length at birth for the Bundi were obtained by extrapolating back from the measurement made at 1 month of age. All three groups are exposed to nutritional and infectious disease stresses.

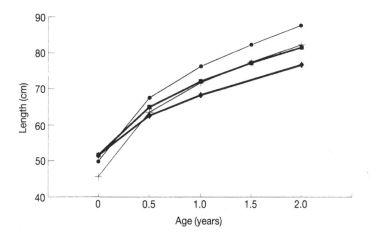

Figure 6.1. Length by age of two rural third-world populations, the Bundi of PNG (◆) (Malcolm, 1970) and the Kasongo of Zaire (■) (van Lerberghe, 1990), compared with the NCHS (1977) 50th (●) and 5th (+) centiles, 0 to 2 years.

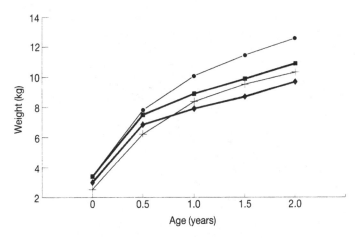

Figure 6.2. Weight by age of two rural third-world populations, the Bundi of PNG (◆) (Malcolm, 1970) and the Kasongo of Zaire (■) (van Lerberghe, 1990), compared with the NCHS (1977) 50th (●) and 5th (+) centiles, 0 to 2 years.

At birth, both the Bundi and Zaire populations have similar mean values to the NCHS references, but by 1 year of age, both weight and length have declined relative to reference values. This pattern is fairly typical for non-industrialised populations living in poor-quality environments. It

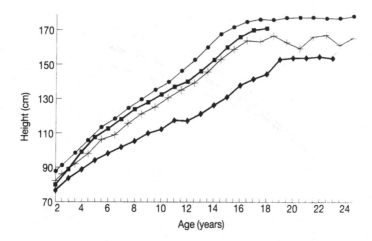

Figure 6.3. Height by age of two rural third-world populations, the Bundi of PNG (♦) (Malcolm, 1970) and a group in Somalia (■) (Gallo & Mestriner, 1980), compared with the NCHS (1977) 50th (●) and 5th (+) centiles, 2 to 24 years.

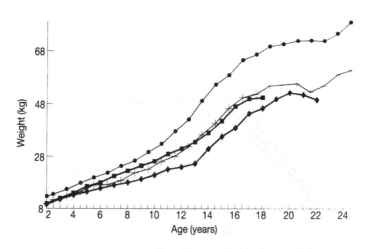

Figure 6.4. Weight by age of two rural third-world populations, the Bundi of PNG (♦) (Malcolm, 1970) and a group in Somalia (■) (Gallo & Mestriner, 1980), compared with the NCHS (1977) 50th (●) and 5th (+) centiles, 2 to 24 years.

might be expected that Somali children would exhibit a similar pattern; certainly, at age 2 years, mean heights and weights of Kasongo (Zaire) and Somali boys are very similar. Between the ages of 2 and 24 years, the growth patterns of Somali and Bundi children differ. The Somalis appear to follow the NCHS pattern, albeit at a lower level, experiencing the adolescent growth spurt at a similar time to the North American reference population, while the Bundi children continue to deviate from the NCHS references more-or-less throughout their childhood, with a delayed adolescent growth spurt. Assuming that the growth pattern of Somali children before 2 years of age is similar to that of Kasongo children, Figs. 6.1–6.4 suggest that the growth of the Somali children may have reached accommodation with the availability of nutritional resources by 2 years of age, while the growth of the Bundi children has not. The Bundi continue to become small across subsequent childhood and adolescence. This illustrates the two extremes of the most typical growth patterns of populations in developing countries, where the process of becoming small is either (1) curtailed in infancy, once body size for age matches food availability for the maintenance of that body size, or (2) continues through childhood until the energy cost of supporting the size of the body matches food availability. At the most extreme, the process of becoming small can continue throughout puberty, resulting in delayed onset of puberty and attainment of final adult body size.

The smaller body size of adults in the developing world relative to the industrialised nations is largely a result of a process of accommodation or adaptation to nutritional availability and, in earlier life, to exposure to infectious disease. Once an individual has gone through the entire period of growth and development and epiphyseal fusion has taken place, skeletal size is set for the rest of life. However, prior to skeletal maturity, it is possible to regain lost growth if conditions improve. The phenomenon of catch-up growth was first described by Prader *et al.* (1963) and has been observed by many since in the context of nutritional rehabilitation. On the basis of the observation that height delay is greater than delay in skeletal maturation in stunted children, Martorell *et al.* (1979) suggested that catch-up growth in chronically malnourished children is limited and related to maturity delays. This and subsequent formulations (Martorell, Rivera & Kaplowitz, 1990; Martorell *et al.*, 1992) suggested that malnutrition in the first 2 to 3 years of post-natal life irrevocably changes the child, so that they are 'locked into' a lower growth trajectory, with lower potential for future growth.

In practice, catch-up growth is not observed outside of the clinical context, since the improvements in the nutritional environment to allow catch-up growth do not occur in nature. However, this does not discount the possibility of catch-up taking place at some stage in childhood. Golden (1994) has examined a number of data sources that suggest that catch-up growth can take place at any time across the course of growth and

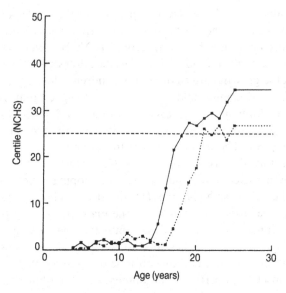

Figure 6.5.   Heights of slaves transported by sea from one part of the USA to another between 1820 and 1860, by age. The data are plotted as centiles of the NCHS (1977) reference. The horizontal line is at the 25th centile. —, Male; ...., Female. Data from Steckel (1987). (From Golden, 1994.)

development. These include: (1) comparisons of bone age and height age; (2) follow-up of malnourished children; (3) immigrant studies; (4) adoption studies; (5) a study of catch-up growth in formerly abused children (King & Taitz, 1985); (6) catch-up after secondary malnutrition associated with renal and intestinal disease; and (7) an historical study of catch-up growth among slaves in the USA (Steckel, 1987). He concludes that with a change in the environment, through adoption, emigration or treatment of a disease associated with growth faltering, catch-up growth occurs, but often is not complete. However, if puberty is delayed and/or growth continues into the early or mid-twenties, then final height may be close to complete. In addition, evidence of almost complete catch-up comes from the study of growth depression and recovery in slaves in the USA (Fig. 6.5).

In a more limited review, Martorell, Khan and Schroeder (1994) have moved their position away from that of the 'locked trajectory', and closer to that of Golden (1994), when they say that prolongation of the growth period can make up for slow growth in earlier life and that in settings where maturation is grossly delayed relative to western references the potential for catch-up in growth can be marked. However, they are more cautious than Golden when they say that growth faltering in school age children and

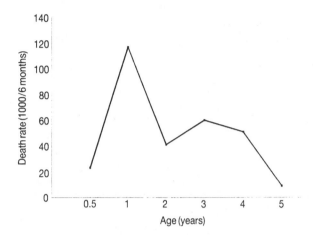

Figure 6.6. Child mortality in Keneba, the Gambia, 1949–53 (from McGregor *et al.*, 1961).

in adolescents may be less reversible than growth faltering in early childhood (Martorell *et al.*, 1994). Both Golden (1994) and Martorell *et al.* (1994) agree that subjects who stay in the setting in which the growth faltering took place experience little or no catch-up growth later in life.

One of the costs of 'becoming small' in infancy is the high mortality associated with it. Child mortality in Keneba, the Gambia was estimated during the years 1949–53, when the availability of western curative medical services in this village were low (McGregor, Billewicz & Thomson, 1961) (Fig. 6.6). Under conditions of low medical intervention, the peak in child death rate was at 1 year of age, with child survivors aged 8 to 10 months being on average 0.5 kg heavier and 3 cm taller than their age-mates who died. Furthermore, various studies of prospective mortality have been carried out in relation to anthropometric status and have shown greater risk of death in children with either low weight for age, or low height for age (Pelletier, 1991) (Tables 6.1 and 6.2).

After infancy, the effects of infection on physical growth are small (Rowland, Rowland & Cole, 1988), largely because of the development of immune system memory to the local disease ecology. Having become small, low total food availability usually denies the possibility of catch-up growth.

The energy cost of growth and bodily maintenance

The energy cost of physical growth and development has been assessed by totalling the energy cost incurred in deposition of new tissue to an infant or

Table 6.1. *The relationship between child mortality and weight for age as a percentage of international median*

Country	Child mortality (deaths/1000 per year) at weight for age (% of median)				Reference
	>80	70–80	60–70	<60	
Tanzania	20	50	70	210	Yambi, 1988
PNG	25	50	60	150	Heywood, 1982
Bangladesh	50	20	50	100	Coghill, 1982
Bangladesh	0	10	20	60	Alam *et al.*, 1989
Bangladesh	0	10	20	50	Bairagi *et al.*, 1985
India	0	5	10	40	Kielmann & McCord, 1978

Source: From Pelletier (1991).

Table 6.2. *The relationship between child mortality and height for age as a percentage of international median*

Country	Child mortality (deaths/1000 per year) at height for age (% of median)				Reference
	>95	90–95	85–90	<85	
Tanzania	30	25	67	56	Yambi, 1988
PNG	18	25	35	72	Heywood, 1982
Bangladesh	35	40	38	115	Coghill, 1982
Bangladesh		7	10	30	Alam *et al.*, 1989
Indonesia	22	30	38	38	Katz *et al.*, 1989

Source: From Pelletier (1991).

child, and this figure has been related to measures of energy expended in maintenance metabolism and physical activity. The accuracy of this work depends on the precision with which body composition, energy balance and energy expenditure can be estimated in infants and children. Advances in methodology in these areas have been considerable in recent times. Figure 6.7 shows the energy cost of growth, in addition to the energy expended in both maintenance metabolism and physical activity, for Gambian and British children combined (Prentice *et al.*, 1990). This is based on measurements of energy expenditure and calculated estimates of the energy cost of tissue deposition from body composition estimates. Estimation of the energy cost of growth as a proportion of TEE by a different method is given in Fig. 6.8. This is based on estimates for an hypothetical child born at, and following, the 50th centile of the NCHS references of weight for age

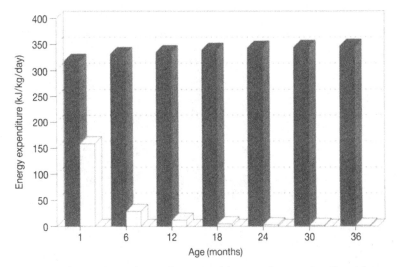

Figure 6.7. Estimate of energy requirements for growth (□) and for maintenance and physical activity combined (■) in children aged 1 to 36 months (from Prentice *et al.*, 1990).

(National Center for Health Statistics, 1977). In calculating these figures, reported values for sleeping metabolic rate (SMR), an approximation of BMR used in situations where it is impossible to measure BMR (Vasquez-Velasquez, 1988), and the energy cost of tissue deposition (Bergmann & Bergmann, 1986) at different ages in infancy are used. In addition, it is assumed that the energy cost of activity for an individual is 40% above BMR (FAO/WHO/UNU, 1985).

Figure 6.7 shows that the energy cost of growth is high, but only in the earliest stages of post-natal life. At 1 month of age, this is about one third of total energy expenditure, declining to about 4% by 1 year of age. By 2 years of age, the energy cost of growth is a mere 1% of total energy expenditure. Figure 6.8 is in broad agreement with Fig. 6.7, showing the energy cost of growth to be 25% of estimated TEE at 1–3 months of age, declining to 4% of TEE at 9–12 months of age, 2% at 18–24 months of age. Since the energy cost of growth is low beyond about the first year of life, it seems unlikely that underfeeding is the primary cause of growth faltering beyond this age. Energy expended in maintenance metabolism is overridingly the most important component of TEE and, therefore, of dietary energy require-ments, while the energy expended in physical activity must be important in cognitive and behavioural development, particularly when the child becomes more mobile. The process of becoming small must be considered

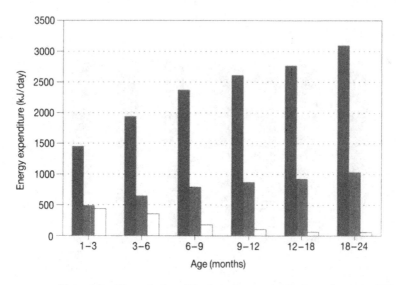

Figure 6.8.   Theoretical partitioning of energy intake for maintenance (■),
growth (□) and physical activity (■) of a 50th centile (NCHS, 1977) child
from birth to 2 years of age (from Ulijaszek & Strickland, 1993a).

in relation to the supply of dietary energy for the high cost of bodily
maintenance and of defending attained body size, and the behavioural
consequences of reduced physical activity in the face of reduced food
supplies relative to desirable levels of output. More importantly, the
process of becoming small must be viewed as an accommodation to both
food availability and the local disease ecology, factors which interact. In
particular, poor nutritional status can impair immune function and
increase susceptibility to infectious disease (Ulijaszek, 1990b), while
episodes of infectious disease can lead to reduced food intake, impaired
absorption of energy and nutrients, increased metabolic costs and altered
susceptibility to future disease challenge (Tomkins & Watson, 1989).

Nutrition and infection

During the first 3 months of life, breastmilk output of the mother is
generally able to sustain the energy demands of the child (Whitehead,
1985). There is also passive immunity obtained from the mother at this time
(Rahaman *et al.*, 1987). Adequate nutrition and passive immunity usually
minimise the impact of infection on the growth of exclusively breastfed
young infants (Zumrawi, Dimond & Waterlow, 1987; Rowland *et al.*,
1988). With the introduction of supplementary food, the child's potential
exposure to pathogenic agents increases and with it comes an increased risk

Table 6.3. *Prevalent diseases in Africa, Asia and Latin America*

	Infections (1000 cases/year)	Deaths (1000 cases/year)
Diarrhoeas	3–5 000 000	5–10 000
Ascariasis	8–1 000 000	20
Tuberculosis	1 000 000	400
Hookworm	7–9 000	50–60
Malaria	800 000	1200
Trichuriasis	500 000	Low
Amoebiasis	400 000	30
Filiariasis	250 000	Low
Giardiasis	200 000	Very low
Schistosomiasis	200 000	500–1000
Measles	85 000	900
Polio	80 000	10–20
Whooping cough	70 000	250–450
Onchocerciasis	30 000	20–50

Source: From Walsh & Warren (1979).

of infection. If the introduction of supplementary foods is delayed, then the energy costs of growth are not met, depression of immune responsiveness takes place and again risk of infection is increased (Tomkins & Watson, 1989).

Factors that are important in determining energy and nutrient intakes and patterns of infectious disease in young children include: (1) geographical and climatic conditions that permit or prevent the possibility of pathogens thriving; (2) local behaviours and cultural factors that serve to make a disease endemic in any population; (3) infant-feeding practices; (4) the nature of the local diet and the foods offered to children; (5) environmental sanitation; and (6) the degree of contamination of foods and liquids. Some of these factors may vary with season and the age of the child (Mata, 1983). For example, tropical climatic seasonality, living in confined spaces and at high population density, and environmental sanitation practices often combine to make the incidence of diarrhoea seasonal, but variable from population to population. Younger children may be more susceptible to some types of disease than are older children, because they have not yet acquired immunity.

Of all infectious diseases, diarrhoeas caused by bacterial or viral infection have the highest prevalence in the developing world, followed by infections caused by intestinal parasites, tuberculosis and malaria (Table 6.3). They are also the most important non-dietary cause of growth faltering in young children in developing countries (Rowland & Rowland, 1986).

The relationships between energetics, infectious disease and physical growth in infants and young children has been extensively studied in both rural and urban Gambia (Tomkins, 1986b; Rowland *et al.*, 1988). Diarrhoea and upper respiratory tract infection (URTI) are of equal incidence in children below the age of 2 years. For URTI, the proportion of all days spent ill falls from 28% in children below 6 months of age, to 13% in those above 18 months of age. Incidence of diarrhoea rises from 16% of all days ill below 6 months of age, to 37%, above 18 months of age. Lower respiratory tract infection (LRTI) has a much lower incidence: 4% of all time spent ill is the result of LRTI in all children between 0–24 months of age, with little variation across age groups (Rowland *et al.*, 1988).

Authors have collected data and information on a range of topics, including: (1) body weight (Ann Prentice, Cole & Whitehead, 1987a); (2) weight gain (Waterlow, Ashworth & Griffiths, 1980); (3) SMR (a proxy for BMR, which is impossible to measure under the conditions prescribed by Boothby and Sandiford (1929)) (Vasquez-Velasquez, 1988); (4) TEE from doubly labelled water measurements (Vasquez-Velasquez, 1989); (5) breastmilk consumption (Ann Prentice, 1986); (6) breastfeeding practice and supplementary feeding (Barrell & Rowland, 1979; Barrell & Kolley, 1982; Whitehead, 1985); (7) immune status of infants and young children (McGregor *et al.*, 1961; Ann Prentice *et al.*, 1984); and (8) effects of diarrhoea on food intake and absorption (Molla, Molla & Khatun, 1986).

Table 6.4 gives measures of: (1) energy expenditure partitioned into maintenance, activity, and growth (Vasquez-Velasquez, 1988); (2) weight gain (Ann Prentice *et al.*, 1987a); and (3) estimates of the proportion of TEE expended in growth and physical activity for Gambian children aged 0–18 months. Figure 6.9 shows Z scores for weight for age (W/A) relative to NCHS (1977) references.

In the first 3 months of life, weight gain is comparable to that exhibited by children in Britain and the USA (Offringa & Boersma, 1987), despite body weight being around -1 Z score of the NCHS (1977) median at this time (Fig. 6.9). Breastmilk intake alone is adequate to sustain this level of weight gain. Dietary supplementation generally begins when the infant is between 3 and 5 months of age (Whitehead, 1985), at a time when immunity, as measured by serum gamma-globulin levels, is at its lowest and the breastmilk concentration and the absolute intake by infants of IgA, IgG, IgM, C3, C4, lactoferrin and secretory component is declining (Ann Prentice *et al.*, 1984). Supplementation takes the form of millet gruel, often supplemented with sour cow's milk (Barrell & Kolley, 1982), both foods

Table 6.4. *Estimated partitioning of energy intakes of Gambian children aged 0 to 18 months[a]*

Age group (months)	Weight gain[b] (g/day)	Energy expenditure kJ/day)[c]			Energy available for	
		SMR[d]	Growth	Activity	Growth (%)	Activity (%)
0–1	24.1	860	335	430	21	26
1–3	25.6	1134	551	401	26	19
3–6	12.9	1534	248	669	10	27
6–9	8.2	1683	103	774	4	30
9–12	6.8	1754	74	920	3	33
12–18	6.5	1888	68	942	2	33

[a] From Ulijaszek & Strickland (1993a).
[b] From Ann Prentice *et al.* (1987a).
[c] From Vasquez-Velasquez (1988).
[d] Sleeping metabolic rate as a measure of maintenance energy requirements.

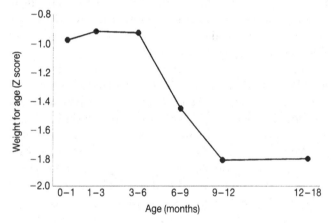

Figure 6.9.   Weight for age (Z scores of NCHS, 1977) for Gambian children between the ages of 0 and 24 months (from Ulijaszek & Strickland, 1993a).

containing high levels of coliform bacteria capable of inducing diarrhoea (Barrell & Rowland, 1979). The prevalence of diarrhoeal disease increases from the age of 3 months, peaking at 9 months of age (Rowland, Cole & Whitehead, 1977).

In the age range 3 to 12 months, the proportion of TEE available for growth is lower than values presented by Prentice *et al.* (1990) (Fig. 6.7), and the estimate of energy cost of growth of an hypothetical NCHS 50th centile child (Fig. 6.8). Weight gains relative to references are also lowest in this age range; mean W/A is about −0.9 Z scores of the NCHS median in the age groups 1–3 and 3–6 months, falling to about -1.8 Z scores in the age group 9–12 months and remaining at this level in the 12–18 months age group. In the 9–12 months age group, the low W/A may reflect poor current nutritional status. In the 12–18 months group, it is more likely to be an outcome of growth faltering at an earlier age and may not reflect current undernutrition; this would be the case if decline in weight were followed by faltering in length or height relative to growth references, the process of 'becoming small' having been completed.

In the age groups where growth faltering takes place (6–12 months), the energy available for physical activity is many times greater than that available for growth. For example, in the age group 6 to 9 months, the energy expended in growth is 4% of TEE, while the energy expended in activity is 30% of TEE. In the 9–12 months age group, energy expended in growth is 3% of TEE, while energy expended in physical activity is 33% of TEE. Given that the incremental energetic cost of sustaining a level of growth similar to that prescribed by the NCHS (1977) references is a mere 2–3% of TEE in addition to that already committed to growth, simple energy deficiency cannot be invoked as the explanation for the process of becoming small in these age groups and beyond.

The expenditure of a high proportion of TEE in physical activity at a time of life when W/A is declining could be related to the activity requirements for adequate behavioural and cognitive development. Below 3 months of age, children's ability to perform gross motor tasks is limited, and cognitive development and exploratory behaviour requires little physical exertion. After 3 months of age, gross motor abilities increase rapidly, and with them comes more active exploratory behaviour. However, reduced energy intake results in reduced physical activity, even in children with mild to moderate malnutrition (Grantham-McGregor *et al.*, 1990), supporting the view that limited energy intake is unlikely to be a primary factor in limiting the growth of Gambian children.

The most likely explanation is that a shortage of dietary factors other than energy could be implicated in the growth-faltering process. This could involve deficiencies of other dietary factors such as protein (Golden, 1988), zinc (Golden, 1988; Ann Prentice & Bates, 1994), calcium (Fraser, 1988; Ann Prentice & Bates, 1994), or even sulphur (Golden, 1994). Indeed it has been suggested that the diets of Gambian children might be deficient in calcium, but not zinc (Ann Prentice & Paul, 1990). However, the impact of repeated episodes of infections such as diarrhoea, URTI and LRTI should not be under-estimated.

The impact of diarrhoea on growth results from several factors, of which the decreased intake of food caused by anorexia or the withholding of food

Table 6.5. *Intake of dietary energy by infants at different stages of diar-rhoeal infection, by type of diarrhoeal infection*

	Dietary energy intake (kJ/kg per day)[a]		
	Acute stage	2 weeks after infection	After recovery
Cholera	313 (151)	465 (148)	458 (133)
Enterotoxigenic *Escherischia coli*	296 (159)	381 (119)	471 (79)
Shigella sp.	293 (118)	420 (116)	457 (79)
Rotavirus	287 (95)	365 (110)	481 (84)

Source: From Molla *et al.* (1982).
[a] Dietary energy intake as mean with standard deviation in parentheses.

Table 6.6. *Total energy intake from breastmilk, weaning foods and oral fluids in children with diar-rhoea and in healthy controls, in Bangladesh*

	Energy intake (kJ/kg per day)[a]	
	Healthy controls	Children with diarrhoea
Sample size	11	15
Source of intake:		
Breastmilk	224 (93)	196 (89)
Oral fluids	–	42 (32)
Weaning foods	361 (293)	139 (143)
Total energy intake	585 (233)	314 (134)

Source: From Hoyle *et al.* (1980).
[a] Energy intakes given as means with the standard deviation in parentheses.

is often the most important (Molla *et al.*, 1986). Table 6.5 gives mean intake of dietary energy in Bangladeshi infants according to the type of diarrhoea, while Table 6.6 gives the energy intake from different sources in Bangladeshi children with diarrhoea and in healthy controls. According to these data, there is clear suppression of energy intake in the acute stage of diarrhoeal infection, with the increase in intake 2 weeks after infection varying according to the type of diarrhoeal infection. Comparing energy intakes at 2 weeks of infection and after recovery with intakes in the acute stage of infection, the most rapid recovery of intake occurs with cholera

Table 6.7. *Expected weight changes caused by diarrhoea in Gambian children at different ages*[a]

Age group (months)	Weight[b] (kg)	Maximum energy intake[c] (kJ/day)	Energy expenditure Maintenance[d] (kJ/day)	Growth[e] 9kJ/day)	Change in body mass (g/day)	(g/kg per day)
0–1	3.16	1624	1204	+420	+30.2	+9.6
1–3	4.31	2085	1588	+497	+23.1	+5.4
3–6	6.11	1792	2148	−356	−18.5	−3.0
6–9	7.07	2050	2356	−306	+24.3	−3.4
9–12	7.76	2250	2456	−206	−18.9	−2.4
12–18	8.66	2511	2643	−132	−12.6	−1.4

[a] From Ulijaszek & Strickland (1993a).
[b] From Waterlow *et al.* (1980).
[c] Breastmilk intake values from Ann Prentice (1986). Maximum energy intakes calculated assuming that breastmilk intake is not impaired by the onset of diarrhoea and that energy intake of children receiving a predominantly solid diet is 290 kJ/kg body weight.
[d] $1.4 \times$ SMR.
[e] Difference between maximum energy intake and maintenance energy expenditure; energy cost of tissue deposition from Bergmann and Bergmann, 1986.

infection, while the most gradual recovery is with rotaviral and enterotoxigenic *Escherischia coli* infection. Suppression of intake during diarrhoea comes almost totally from reduced intake of weaning foods.

During severe diarrhoea, energy intakes of children receiving a predominantly solid diet may be only 290 kJ/kg body weight per day (Molla *et al.*, 1982; Tomkins, 1983), while infants receiving predominantly breastmilk show no decline in energy intake under similar conditions (Hoyle, Yunus & Chen, 1980; Brown *et al.*, 1985). Assuming these levels of intake and that habitual levels of breastmilk intake are not impaired by the onset of infection, expected changes in body weight (g/kg per day) caused by the onset of severe diarrhoea have been calculated for Gambian children (Table 6.7) (Ulijaszek & Strickland, 1993a).

If breastmilk intake is not impaired by the onset of diarrhoea, then there is no effect on the weight gain of infants below the age of 3 months. The effect of diarrhoea-induced food restriction or anorexia on weight change appears to come between the ages of 3 and 9 months, when the energy cost of growth is still high, but it persists at a lower level up to the age of 18 months, and possibly beyond. Given that the incidence of diarrhoea in Gambian children differs little between age groups below the age of 18 months (Rowland *et al.*, 1988), it is possible that the weight loss illustrated in Fig. 6.9 may be caused in large part by the low energy intakes associated

Table 6.8. *The impact of disease on weight and length gain in Gambian infants*

	Weight (g/day)		Length (mm/day)	
	Regression coefficient	SEM	Regression coefficient	SEM
Upper respiratory tract infection	−2.3	1.8	0.00	0.05
Lower respiratory tract infection	−14.7	3.1	−0.04	0.09
Diarrhoea	−3.7	1.2	0.01	0.04

Source: From Rowland *et al.* (1988).
Sample size: 126 children; SEM, standard error of the mean.

with diarrhoeal infection at the time of infection, rather than overall restricted or low food availability.

There is considerable variation in the impact of disease on length gain in Gambian infants, as illustrated by very large standard errors around regression coefficients for length gain against URTI, LRTI and diarrhoea (Table 6.8). Although the effect of LRTI on weight gain is much greater than the effect of diarrhoea, the prevalence of LRTI is a fraction of that of diarrhoea. This results in an overall effect on weight gain in the infant population that is about half that of diarrhoea (Rowland *et al.*, 1988). Some of the variation in disease impact on length gain may be caused by differences in the way that infection in a child is managed, and the duration and intensity of infection. However, another possibility exists. Lampl (1993), carrying out daily measurements of length in a group of healthy North American children, demonstrated a process of incremental growth in length in the first 2 years of life that is characterised by discrete episodic events of amplitudes of 0.5–2.0 cm growth taking place over periods of 1–7 days, separated by periods of no measureable growth of 2–60 days. These growth saltuses do not appear to be cyclical (Lampl, Veldhuis & Johnson, 1992) and show enormous within- and between-subject variation (Lampl, 1993). Saltatory growth has yet to be demonstrated in infants in developing countries, although phenomena described as alternating catch-up and catch-down growth have been described for infants in Ethiopia (Harrison *et al.*, 1990) and Sudan (Brush *et al.*, 1992).

If such bursts of growth represent a normal pattern of growth in Gambian children, then the impact of disease on growth would be greater if it coincided with the time of a growth saltus. It is not clear whether the timing and extent of growth saltus is environmentally labile; however, the most likely way in which infection such as diarrhoea might interfere with saltus is through reduced food intake, either because of anorexia or food restriction by the mother. Reduced intake of dietary energy and protein could inhibit or switch off growth saltus through reduced activity of insulin-like growth factor I (IGF-I) and the principal thyroid hormone, triiodothyronine (T_3).

Energy and protein intakes are important factors in the regulation of serum IGF-I (Isley, Underwood & Clemmons, 1983; Underwood et al., 1989). IGF-I has bone as a primary target, and this is perhaps the most sensitive pathway by which saltus might be inhibited by reduced food intake. During fasting, serum IGF-I may drop to 40% of usual, non-fasting levels by 5 days; the fall in T_3 in response to low food intake occurs within 24 hours, reaching values 40% lower than in normally fed subjects within 48–72 hours (Danforth, 1985). At the cellular level, the action of T_3 may involve two interrelated processes. The first is its effect on metabolic rate and consequent energy balance. The second is on growth and development, through its influence on gene expression, protein synthesis and protein turnover (Millward, 1986). In the pituitary, thyroid hormone regulates the synthesis of growth hormone (Danforth & Burger, 1989), which is released in pulsatile fashion (Hartman et al., 1993). Although growth in the first 6 months of post-natal life may involve a growth hormone-free mechanism (Karlberg & Albertsson-Wikland, 1988), this does not extend to the period in which growth faltering owing to diarrhoeal infection in children in the developing world is most common: beyond 6 months of age.

Undernourished children can conserve nitrogen from protein when their energy intakes fall (Jackson, 1985) and rates of whole-body protein turnover are depressed (Golden, Waterlow & Picou, 1977). However, during growth, a fall in T_3 has a greater inhibitory effect on protein synthesis than on degradation: inhibiting growth rather than sparing it (Millward, 1986).

The effects of nutrition and infection on the growth of children between the age of 3 years and the onset of puberty are smaller than at the earliest ages but are related to overall energy resources. If adjustment to resources does not take place in the first few years of life, then it may continue at a lower level across childhood and, at the most extreme, into puberty. Although growth in adolescence has a much larger genetic component than growth in pre-adolescence (Johnston et al., 1976), the timing of the onset of puberty and the size of the growth spurt are affected by environmental factors (Eveleth & Tanner, 1990).

Adult body size

What are the biological implications of small body size in adulthood? Various studies have shown relationships between body size and physical

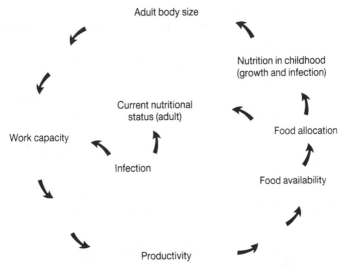

Figure 6.10. The energy trap model.

work capacity ($\dot{V}O_2$max) (Ferro-Luzzi, 1985; Spurr, 1988a) and work productivity (Satyanarayana *et al.*, 1977; Spurr *et al.*, 1977; Brookes *et al.*, 1979; Immink & Viteri, 1981; Immink *et al.*, 1984; Spurr, 1988b). All of these studies examined these relationships in situations where adult males were engaged in paid labour. However, it is not clear whether similar relationships exist in populations occupied primarily in traditional subsistence economies (Strickland, 1990).

Understanding such relationships in traditional subsistence systems is difficult because of great within- and between-community variation in: (1) intensity and duration of work; (2) day-to-day work patterns; (3) patterns of group work according to task and season; and (4) direct seasonal and climatic influences on work performed. Although complex, a general model in which the relationships between growth, body size, work capacity and subsistence productivity might operate in such contexts is summarised in Fig. 6.10.

This is an elaboration of an earlier model proposed by Martorell and Arroyave (1988) and has been used to test the proposition that small body size is adaptive for one New Guinea group of hunter–horticulturalists, the Wopkaimin (also described in Chapter 2, in a related analysis). Briefly, a household, family or unit of reproduction can be considered to be 'energy trapped' if their rate of energy flow and productivity is so low that some aspects of biological function, including reproductive success, are impaired. In this analysis, adult Wopkaimin males were divided into three categories, tall, short, and short and thin, and their ability to perform

Table 6.9. *Testing the energy trap model: adult male body size, work potential and mortality and body size of children, Wopkaimin, Papua New Guinea*

	Adult body size		
	Tall	Short	Short and thin
Sample size	10	21	5
BMI (kg/m²)	> 18.5	> 18.5	< 18.5
Height (cm)	> 164	< 164	< 164
$\dot{V}O_2$max (l/min)	2.43	2.01	1.80
$\dot{V}O_2$max (ml/kg per min)	39.1	38.4	39.4
Proportion of $\dot{V}O_2$max spent in key activity, clearing ground (%)	37	43	49
Energy intake (MJ)	8.81	8.62	7.46
Duration of daily activity which could be sustained at existing levels of habitual energy intake (hours)	6.4	8.4	3.1
Number of offspring in family	2.7	2.6	3.0
Mortality rate in children under 5 years (per 1000 live births)	207	214	286
Proportion of children in family with BMI below -2 Z scores of NCHS median (%)	11	6	40

physical work was examined, using measures of oxygen consumption at different levels of activity (Brown, 1987). This was then related to the energy cost of subsistence tasks, as measured by Norgan *et al.* (1974). Subsequently, the energy cost of subsistence work at the three different body sizes was related to habitual energy intakes, measured by Ulijaszek (1987), with the aim of examining the duration of work output capable of being sustained by tall, short, and short and thin men. Finally, the family size, mortality of young children and body size of survivors are given for families whose head of household is either tall, short, or short and thin. Results are given in Table 6.9.

Differences in $\dot{V}O_2$max between the three groups disappear when the variable is expressed per unit of body weight, although the proportion of $\dot{V}O_2$max expended in one key activity, garden clearing, varied between the three, being greatest in short and thin men. For both short, and short and thin men the proportion of $\dot{V}O_2$max expended in this activity exceeds 40%, suggesting that such work could not be maintained continuously for an 8-hour period in the course of a day (Spurr, 1988a). The energy cost of work also varies with body size, and it is possible that a small body size requires less energy for subsistence activity. Habitual levels of energy intake vary between the three groups of men, being lowest for the short and thin men. However, the difference in the number of hours of subsistence work sustainable at these levels of intake is quite large. Although tall men have greater energy intakes than short men, they can only sustain 6 hours of hard physical work in the course of a day, while short men can sustain more than 8 hours of work. This analysis suggests that short and thin men would only be able to work for a maximum of 3 hours without going into negative energy balance.

Although caution is advised in the use and interpretation of energy-intake data, if it can be assumed that any bias in intake or reporting are similar for all Wopkaimin studied, body size differences appear to have some important ecological correlates. Short men appear to be 'energy trapped' by their low $\dot{V}O_2$max, while short and thin men appear to be trapped by both their low $\dot{V}O_2$max and their habitual energy intakes. Since the Wopkaimin have no need to sustain high levels of work output for more than 6 hours on a regular basis, it is unlikely that tall men may also be energy trapped. Functional consequences of energy entrapment are evident in the offspring of these three groups of men. Children of the short and thin men have higher mortality rates and smaller BMI than children of either short, or tall men.

This example illustrates the use of an ecological 'energy trap' model in examining body size-related energy constraints within a subsistence economy; the model could be used to consider other relationships, such as (1) within-household food allocation on body size of adults and the growth of children; (2) seasonality on work capacity, productivity and food availability; or (3) effects of infection on work capacity and productivity.

Summary

In this chapter, the importance of energetics to growth and bodily maintenance and the functional implications of small body size in children and adults are examined. The smaller body size of adults in the developing world relative to the industrialised nations is largely a result of a process of accommodation or adaptation to nutritional availability and exposure to infectious disease. Although various sources of evidence suggest that catch-up growth can take place at any time across the course of growth and development (Golden, 1994), this rarely happens in practice, since subjects who stay in the setting in which the growth faltering took place experience little or no catch-up growth later in life.

The energy cost of growth is high only in the earliest stages of post-natal life. At 1 month of age, this amounts to about one third of total energy expenditure, declining to about 4% by 1 year of age. By 2 years of age, the energy cost of growth is a mere 1% of total energy expenditure. Since the energy cost of growth is low beyond about the first year of life, it seems unlikely that underfeeding is the primary cause of growth faltering beyond this age. Rather, the interactive effects of infectious disease and undernutrition on growth are more important. Poor nutritional status can impair immune function and increase susceptibility to infectious disease, while episodes of infectious disease can lead to reduced food intake, impaired absorption of energy and nutrients, increased metabolic costs and altered susceptibility to future disease challenge.

Various studies have shown relationships between body size, $\dot{V}O_2$max and productivity in adults engaged in paid labour. Similar relationships are shown for a group of hunter–horticulturalists in Papua New Guinea, the Wopkaimin. Short men appear to be 'energy trapped' by their low physical work capacity, while men who are short and thin appear to be trapped by both their low physical work capacity and their habitual energy intakes. Since the Wopkaimin have no need to sustain high levels of work output for more than 6 hours on a regular basis, it is unlikely that tall men may also be energy trapped. Functional consequences of energy entrapment are evident in the offspring of these three groups of men. Children of the short and thin men have higher mortality rate and smaller BMI than children of either short or tall men.

7   *Energy, effort and subsistence*

Subsistence behaviour is influenced by the environment but also shapes the environment to an extent that varies with the level of technological input. Work organisation varies with the type of subsistence practice adopted, and this in turn affects both long-term and short-term work-group organisation. Long-term organisation includes the formation of a community, band or village in which sharing of dietary energy between households is in long-term, but not necessarily short-term, balance. Short-term organisation includes the formation of work groups to perform specific tasks which may vary from day-to-day, or may be seasonal in nature. In this chapter, a number of long-term and short-term subsistence organisational strategies will be described, and their energetic rationale examined.

Subsistence practices as adaptation and adaptability

In traditional subsistence economies, aspects of culture can act as adaptive code and have elements which allow appropriate responses to changing environmental conditions. It is embodied in the practices themselves and the rituals that surround them (Rappaport, 1978), in settlement patterns that reflect subsistence patterns and in kinship and alliances (Hyndman, 1979). The brief sketches of the ecological energetics work of Rappaport (1968) and Lee (1965) given in Chapter 1 show two of the ways in which culture as adaptation have been examined using an ecological framework. Fundamental to the notion of culture as adaptive code is that it forms the basis for short-term strategies and behaviours in response to changing nutritional circumstances that arise out of implicit values which may have evolved over many generations.

In traditional communities, settlement patterns have coevolved with subsistence practices and processes, since resource acquisition, especially food and nutrition, are fundamental to human survival. Environmental variables such as climate and geography may have helped to shape the types of energetic adaptation seen in human populations past and present, but clearly humans can also manipulate the environment in a large number of ways, either in the process of resource acquisition or in the course of pursuing non-resource-related aspects of behaviour. However, patterns of resource use cannot be predicted from the knowledge of environmental factors (Ellen, 1982).

Aspects of the environment that are impossible for humans to manipulate directly are the macro- and micro-environmental cycles of temperature and rainfall that vary across the year and between years. Within climatic bounds, human groups can pursue a range of subsistence options, most coarsely delimited by the terms pastoralism, hunting and gathering, and agriculture. These terms mask the reality that human groups may share similar land, temperature and rainfall characteristics but do very different things for food. For example, the Lese and Efe of Zaire live adjacent to each other, but each has shaped their ecosystem differently, the first by practising slash and burn horticulture, the second by hunting and gathering (Bailey *et al.*, 1993). In this way, different ecosystems are created, and with them, the development of distinct biological and behavioural adaptations by the Lese and Efe, respectively.

The issue is complicated by the fact that most human populations practise a mix of subsistence strategies. Indeed, there are few groups of 'pure' pastoralists, hunter–gatherers or agriculturalists. Most pastoralists either practise limited cultivation during part of the year, or trade for agricultural crops (Thomas, 1976; Little, 1989), while some groups of hunter–gatherers may practise limited horticulture, or have some other access to crop foods (Headland, 1987; Bailey *et al.*, 1989; Milton, 1991). Even simple agriculturalists may practise varying degrees of animal husbandry, and in this sense 'pure' agriculturalists and pastoralists form the extremes of a continuum of mixed agricultural/domesticated animal subsistence strategies.

Within groups or populations, there may be variation in the extent to which one strategy is practised relative to another. This may be a function of ability, for example in hunting or fishing, or may depend on the inheritance of certain environmental givens, such as the type, size and quality of land inherited, or of wealth accrued or inherited. Relative wealth may also be a determinant of the type of technology that individuals or families can employ. Thus, even in rural subsistence-oriented societies, socio-economic factors can shape human biological function, especially nutritional status and disease experience (Pacey & Payne, 1985), and can influence demography through differential mortality.

Subsistence and ritual

Ritual is associated with the subsistence cycle in many rural societies. One of the ways in which ritual plays a role in subsistence strategy is described for the Tsembaga Maring of Papua New Guinea, in Chapter 1. Briefly, in this ritual they engage in cycles of warfare and pig raising which are punctuated by ritual pig-kills involving large numbers of animals and exchange within and between warring groups. The pig-killing ritual is carried out to appease the ancestors but serves to regulate the relationships between people, pigs and gardens, and the competition of different human groups for limited land resources. By regulating the frequency of tribal warfare, it helped in the redistribution of the population over time across available land and between territorial groups (Rappaport, 1968). In this way, ritual for the Maring may serve purposes independent of the subsistence quest, but in the raising of pigs for ceremony, these animals may serve as a finely tuned monitor of ecosystemic stress of one group of Maring against others. Growing pig populations may act as amplifiers of the human population. When pigs begin to seriously degrade the environment, the pig-killing ritual signals resumption of warfare, after which new territorial boundaries are set. In this way, human populations may be balanced against available land resources.

Among the Massa and Mussey of Cameroon, there are specific religious occasions related to sorghum cultivation, specifically asking for the rains and for good crops (de Garine, 1993). For the Nuer agropastoralists, ceremonies are made to coincide with the time of plenty that occurs at the transition between the dry season and wet season (Evans-Pritchard, 1940). It has been demonstrated for an agricultural community in the Gambia that where crop storage losses exceed 10%, post-harvest gorging is energetically less wasteful of the agricultural crop than maintaining a steady intake across the year (Dugdale & Payne, 1986). Therefore, ritual gorging may have social purposes, but it may also be the least wasteful energetic strategy.

Settlement patterns

There is a large literature which examines, at least in part, the nature of human settlement patterns and their relation to subsistence (e.g. Chisholm, 1962; Lee, 1969; Stenning, 1971; Nietschmann, 1973; Ellen, 1978; Turnbull, 1974; Ohtsuka, 1983; Ulijaszek & Poraituk, 1983). For hunter–gatherers, resources may be unevenly distributed and usually show seasonal variation in availability; settlement patterns may vary across the year and be determined by resource availability and richness across time. For example, Mbuti pygmies undergo considerable changes in band size and distribution within their hunting territory in the course of a year (Turnbull, 1974), undergoing group fission for several months during the wet season in order to maximise the overall fish catch, and fusion again as environmental conditions change the distribution of resources. However, not all hunter-

Table 7.1. *Animal resource availability to the Tasbapauni Miskito of Nicaragua, according to ecosystem*

Ecosystem	Biomass[a] (kg)	Energy[b] (MJ)	Energy (% of total)
Local			
Beach, lagoon and swamp	3 670	14 130	8
Offshore shallow waters, reefs, cays, deep water	36 780	141 240	74
Distant			
Tropical rainforest and gallery forest	3 570	20 920	11
Pine savannah	2 280	13 360	7

[a] Meat and fish biomass butchered between October 1968 and September 1969. Data from Nietschmann (1973).
[b] Energy values estimated using Food Composition Tables for use in East Asia (Leung, Butrum & Chang, 1972).

gatherer groups need to move location; this is especially so for groups that obtain a greater proportion of their diet from fishing. For example, the Miskito live on the Caribbean coast of Nicaragua and practise fishing and hunting, predominantly from local offshore waters and reefs, but also from local lagoon and swamp ecosystems and more distant pine savannah and tropical rainforest ecosystems (Nietschmann, 1973) (Table 7.1).

Hunting in the tropical rainforest provides only a small proportion of the total energy returns to this community, but this type of appropriation involves travelling great distances to find assured meat-yielding biotopes that have not suffered as a result of general human pressure on resources (Nietschmann, 1973).

The picture is more complicated for hunter–horticulturalists. Such groups are more settled than most groups of hunter–gatherers, and although food yields from cultivated sources are generally greater the closer the area of cultivation is to the focus of settlement, this is not true for non-domesticated food sources (Ellen, 1982). Hunting and collecting takes place mostly in the area circumscribed by the outer limit of cultivation, at least for Nuaulu swidden cultivators in Indonesia (Ellen, 1978), with the geographical distribution of non-domesticated resources being unevenly distributed in both time and space. For the Gidra of Papua New Guinea, the distribution and intensity of space used for horticulture and for hunting show considerable overlap, while the use of sago and coconut palms is less intensive as these resources are only available at greater distance from the settlement (Ohtsuka, 1983) (Fig. 7.1).

Agropastoralist groups may also undergo fission and fusion, dispersion

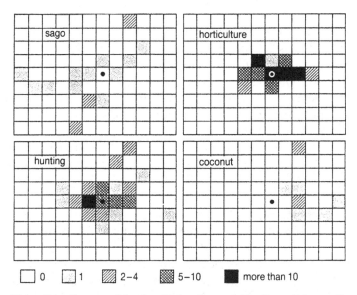

Figure 7.1. Space used for four food-getting activities by an Oriomo village (Gidra) (represented by central dot), PNG. The key indicates the intensity measured as the number of occasions a particular square in the grid was used during a 26-day period of observation (from Ohtsuka, 1983).

and settlement in the course of a year, in response to regular seasonal variations in the demands of the pastoral economy. Such populations generally live in environments where biological productivity is low and is strongly related to rainfall and water run-off (Harris, 1981). The Ferlo of Senegal are a group which show a pattern of subsistence which is typical of many such groups. In the wet season, they aggregate at waterholes in encampments comprising up to five households, each of which contains between eight and ten individuals (Benefice, Chevassus-Agnes & Barral, 1984). This allows them to pool their labour in time-limiting tasks such as the planting and harvesting of millet. At the beginning of the dry season, the encampments disperse, often into smaller units to graze their animals. The resources needed by their cattle are more sparsely distributed across the savannah, and the labour required to maintain them is likely to be small in comparison with the requirements of agriculture. This carries on until the following rainy season, when millet cultivation again calls for the pooling of human physical resources.

The more permanent settlement of a group in one place allows greater food production through the intensification of cultivation and animal husbandry. Centralised patterns of settlement radically alter the relationship between humans and their physical environment, with new emphasis on sustainability of production on land which is not allowed to renew naturally. Intensification of cultivation allowed by centralised settlement patterns involves a shift to the utilisation of only a very limited number of cultigens and changes in social formation. Choice of cultigens may be for greater energy returns, or lower work input. Although this cannot be demonstrated for past populations, it has been suggested that changes in subsistence practices associated with intensification of production among the Baroi of Papua New Guinea have led to the wider adoption of sago cultivars that require lower work input in starch processing (Ulijaszek, 1991).

Subsistence, kinship and alliances

Related to subsistence and settlement patterns are patterns of social relations. For example, a settlement pattern is generated through the interaction of social and ecological processes that arise out of a number of interrelated cycles: (1) the developmental cycle of domestic groups; (2) kin-group fission and fusion; (3) cycles of exchange: ritual and production in general; (4) annual and longer-term cultivation cycles; and (5) the reproductive and growth cycles of animal populations (Ellen, 1982). Of these cycles, the first two are determined by kinship patterns and alliances formed, while the third may be central to group function and involves cooperative behaviour that runs along lines of kinship or alliance. Furthermore, subsistence production may require cooperative behaviour, which, at its simplest, involves a man and wife team, sometimes with the support of children, according to their abilities. More complicated cooperative behaviour includes the formation of male hunting parties and female field labour groups. Cooperative work may permit greater efficiency in the subsistence quest. In cooperative hunting, larger animals may be stalked or the unpredictability of hunting reduced for the individuals taking part; in cooperative labour, tasks which may have a limited window of opportunity may be completed more speedily.

Panter-Brick (1993) has considered the characteristics of different labour strategies, and these are summarised in Table 7.2, while Fig. 7.2 shows the process of decision making between different labour strategies. Of the five categories of labour identified, all apart from hired labour are based on kinship or alliances. If the task is not urgent, or is urgent but not limited by inadequate labour to do the work, then family labour is the preferred option (Chibnick & de Jong, 1989), since this obviates the costs of recruiting and feeding outside workers and ensures a high quality of work performed (Panter-Brick, 1993). Households are bound by ties of mutual

Table 7.2. *Characteristics of different types of labour*

Labour type	Advantages	Disadvantages
Household labour	Low cost; high-quality work; flexible schedule	Long time to complete tasks
Free assistance	Emergency help	Limited duration
Festive labour	Low direct cost	Indirect time and food costs; low-quality work
Reciprocal exchange	Efficiency	Reciprocal obligations
Hired labour	Convenience	High financial cost

Source: From Panter-Brick (1993).

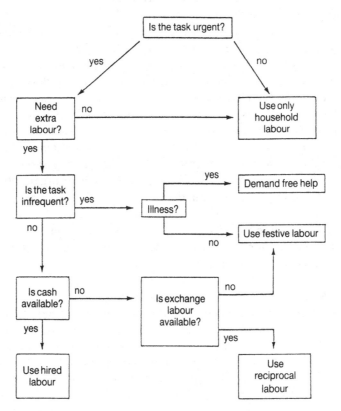

Figure 7.2. Process of decision making between different labour strategies (modified from Chibnick & de Jong, 1989, in Panter-Brick, 1993).

assistance, and, if really needed, free help can be obtained from one's neighbours. Festive labour and reciprocal labour rely on both kinship ties and alliances. Festive labour is organised where a task is of limited duration but occurs annually. The expenditure of cash is avoided, but an outlay of time and resources is needed to prepare food and drink, and the work performed is of low quality (Panter-Brick, 1993). Reciprocal labour, however, is a closely monitored system of exchange favoured where cultivators must complete repetitive tasks at predictable intervals, need work of good quality but cannot afford to hire labour (Panter-Brick, 1993). This involves reciprocal obligations which may be long term.

Subsistence organisation and energetics

The energetics of subsistence organisation can be considered in relation to: (1) the development cycle of domestic groups or households; (2) household fission and fusion; and (3) reciprocal obligations between families and households. Acknowledging that these factors are interrelated, it should be possible to examine: (1) ways in which variations in the developmental cycle of the household can influence household energy requirements and dependency ratios across the lifecycle of the household; (2) whether certain patterns of household development could lead to energetic constraints to the extent that households are faced with a choice of undernutrition, or several options to alleviate this stress, including fission, migration or selling of assets; (3) the extent to which different types of labour organisation are practised according to the stage in the household developmental cycle; and (4) the extent to which reciprocal labour obligations between families and households can serve to reduce energetics constraints at critical stages of the household developmental cycle, and whether labour reciprocity can buffer against early household fission.

Energetics and the household developmental cycle

Although often difficult to define, households have a number of common features which seem to be universal. Simply, households have both a biological and economic base; where systems of inheritance exist, these usually function along household or familial lines (Wilke & Netting, 1984). In many traditional societies, households form the main units of food production and consumption, and they change demographically across the developmental cycle in their overall needs for food and their ability to produce it (Strickland, 1990).

Energy requirements of households across the developmental cycle have been modelled by Ulijaszek and Strickland (1993a), using a set of simplifying assumptions based on typical patterns of child growth and body size at different ages, and on estimates of energy requirements based on body size and activity levels of children and adults (FAO/WHO/UNU, 1985). According to birth interval, household energy requirement peaked between 20 and 30 years after the birth of the first child, the lower value being for a birth interval of 21 months, the higher value for a birth interval of 44 months. The peak energy requirement ranged between about 80 MJ/day for a birth interval of 21 months, and about 55 MJ/day for a birth interval of 44 months. These large differences are caused principally by the difference in total offspring number, the shorter birth interval giving 12 surviving offspring, the longer one giving seven.

Energy-dependency ratios (EDR) based on the assumption that children above the age of 10 years cease to be dependants show clearly that the change of status of children, from dependent to independent or producer, has great implications for the timing of the peak EDR. The model elaborated by Ulijaszek and Strickland (1993a) shows peak EDR at 10 years. If it had been assumed that children ceased to be dependants later, the peak would have shifted to a later stage of household development. However, the model clearly shows a large difference in peak dependency ratio between the household in which the birth interval was 21 months, and the household with a birth interval of 44 months. In what follows, the relationship between the number of offspring, the age at which children cease to be dependants and the EDR across the lifespan of the household is considered for rural Gambian farmers.

Table 7.3 gives mean body weights by age for this population, as well as desirable PALs recommended by FAO/WHO/UNU (1985). Energy requirements were calculated from body weight and PAL, using the Schofield (1985) equations to estimate BMR in all age groups except 1–3 years. For this age group, energy requirement was calculated directly from body weight, using the FAO/WHO/UNU (1985) recommendations for the group. For adults, the PAL used was equivalent to moderate work loads, as recommended by FAO/WHO/UNU (1985). The Schofield (1985) equations were chosen in preference to the Henry and Rees (1988) equations developed for tropical populations, since measured BMR values have been shown to be closer to predictions using the former equations (Lawrence, Thongprasert & Durnin, 1988a).

The total number of offspring in any family can have implications for the ratio of dependants to producers at different stages of the lifecycle of the household. To examine the effects of different family size on the ratio of energy requirements of independent household members (those above 20 years of age) to dependants (those below 20 years of age), household energy requirements were modelled for nuclear families with two, four, six and

Table 7.3. *Weights, physical activity level (PAL) and energy requirement by age group of rural Gambians*

Age (years)	Males			Females		
	Weight[a] (kg)	PAL[b] (TEE/BMR)	Energy requirement[c] (MJ/day)	Weight[a] (kg)	PAL[b] (TEE/BMR)	Energy requirement[c] (MJ/day)
1	10.9		3.7	10.2		3.5
2	14.0		4.9	11.9		4.2
3	15.7	1.64	5.9	14.2	1.64	5.3
4	17.2	1.64	6.1	15.9	1.64	5.6
5	18.5	1.64	6.3	18.0	1.64	5.8
6	19.7	1.64	6.5	19.3	1.64	6.0
7	21.6	1.64	6.8	21.6	1.64	6.3
8	23.8	1.64	7.2	24.5	1.64	6.7
9	27.5	1.64	7.7	27.9	1.64	7.2
10	29.0	1.76	8.0	31.7	1.65	7.7
11	32.4	1.72	8.4	35.7	1.62	8.0
12	36.5	1.69	8.9	40.0	1.60	8.4
13	38.0	1.67	9.1	44.1	1.58	8.8
14	43.3	1.65	9.8	47.8	1.57	9.1
15	48.5	1.62	10.4	50.7	1.54	9.4
16	53.1	1.60	10.8	52.3	1.52	9.6
17	56.3	1.60	11.3	52.8	1.52	9.6
18–30	58.2	1.86	12.2	53.3	1.69	9.0
30–60	57.2	1.86	11.9	52.3	1.69	9.0

[a] Body weight from Billewicz & McGregor (1982).
[b] Desirable PAL from FAO/WHO/UNU (1985).
[c] Energy requirement calculated from body weight and PAL, using appropriate prediction equations of Schofield (1985) in all age groups except 1–3 years. In these groups, it is derived directly from FAO/WHO/UNU (1985) recommendations, per kg body weight.

eight children, respectively, at 5-yearly intervals in households whose lifecycle is 40 years.

The following simplifying assumptions were made:

(i)   Each household consists of one adult male and one adult female and their offspring. There are no elderly dependants or young children from other households.

(ii)   The birth interval is 30 months, with no pregnancy wastage and with alternate male and female births.

(iii)   Dependants are children below the age of 20 years, upon reaching this age they leave the household to form their own.

(iv)   No extra energetic provision is made for pregnancy and lactation.

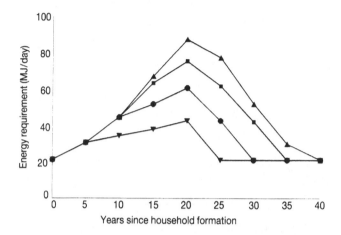

Figure 7.3. Total daily household energy requirements across the lifecycle of households with two (▼), four (●), six (■) and eight (▲) offspring.

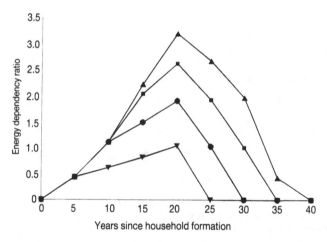

Figure 7.4. Energy-dependency ratios across the lifecycle of households with two (▼), four (●), six (■) and eight (▲) offspring.

(v)   The energy requirements of infants are assumed to be small and are excluded from the analysis.

Figures 7.3 and 7.4 give total daily household energy requirements and EDRs (energy requirements of dependants relative to those of independent members of the household) across the lifecycle of the household. Regard-

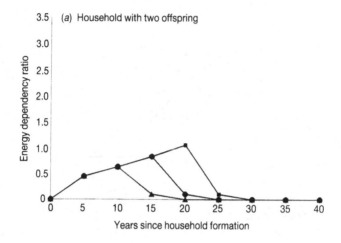

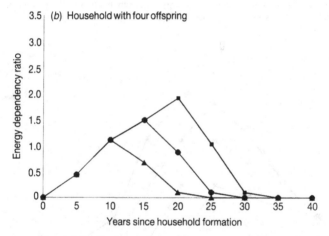

Figure 7.5.   Energy-dependency ratios across the lifecycle of households with offspring becoming independent at 10 (▲), 15 (●) and 20 (■) years of age, in households with: (*a*) two; (*b*) four; (*c*) six; and (*d*) eight offspring.

less of the number of offspring, the peak of household energy requirement and of dependency ratio is at 20 years. This is because of the modelling assumption that children leave the household at this age and are, thus, no longer an energetic burden. However, there are large differences in the peak household energy requirement and EDR according to the number of offspring in the family. The peak energy requirement of the household with eight children is more than twice that of the household with two children; the EDR of the household with eight children is more than three times that

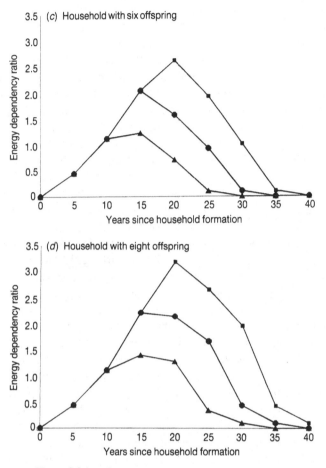

Figure 7.5 (*cont.*)

of the household with two children. It it useful to relate this to the output–input ratio for this group, when practising traditional subsistence. Haswell (1981) estimated that the ratio was 8.6; this can sustain an EDR of about 1.4. In this context, the EDRs do not exceed the capability of agricultural production in the household with two offspring. In households with four, six or eight children, substantial portions of the household lifecycles are spent with energy requirements being in excess of the ability to produce.

One way in which the energetic burden could be reduced is by children contributing to the household economy. However, the extent to which they can be serious contributors to the subsistence economy during late childhood and early puberty is not clear. The effect of children becoming independent members of the household, being neither an energetic burden nor net contributors to the household, is examined in Fig. 7.5.

If children remain dependent until the age of 20 years, then EDRs can only be kept below 1.4 across the household lifecycle if there are two offspring. Larger families cannot sustain adolescent children as economic dependants. If children become economically productive at an earlier age, then larger numbers of offspring can be sustained. For example, if children become economically independent at 15 years of age, then the number of years during which the EDR exceeds 1.4 in households with four, six, and eight children is reduced by at least 5 years. If children become economically productive to the extent of being able to provide the equivalent of their energy needs by the age of 10 years, then households can support up to eight children without exceeding an EDR of 1.4.

In rural Gambia, children begin to assist in agricultural work from about the age of 10 years; about 75% of boys and 50% of girls aged 10 to 16 years in Genieri village were actively engaged in farming (Haswell, 1953). Furthermore, total fertility rates were traditionally above eight (Lunn *et al.*, 1984). This suggests that extensive child labour has been a strategy whereby high numbers of offspring could be supported under ecologically limited circumstances. It also raises the possibility that the decline in fertility rates observed in West Africa (Cohen, 1993) could be related to the withdrawal of child labour in rural subsistence economies, as well as changes in proximate determinants of fertility such as post-partum infecundability and patterns of union (Jolly & Gribble, 1993). Although lower levels of fertility are associated with higher levels of education in mothers (Cohen, 1993), in the rural Gambian context the entry of children into education beyond the primary level may also reduce fertility. This is because it undermines the traditional complex of high fertility, child labour and subsistence labour requirements. According to Fig. 7.5, alternative strategies might be to have four children who cease to be dependent on the household for their dietary energy at 15 years of age, or two children who cease to be dependent at 20 years of age. Thus, the fertility transition in the rural sector of developing countries should be considered in relation to child labour requirements and the effects that increasing the duration of child dependence by sending them to school has on the subsistence economy.

Summary
In this chapter, the adaptive nature of subsistence practices, the organisation of work, settlement patterns, kinship and alliances are examined.

Organisation of work falls within the broader context of group or community organisation and may be long term, or short term. Long-term organisation includes the formation of a community, band or village in which sharing of dietary energy between households is in long-term balance. Short-term organisation includes the formation of work groups to perform specific tasks which may vary from day-to-day, or may be seasonal in nature. Most labour strategies employed by rural populations in the developing world are based on kinship or alliances, but the nature of the strategy employed at any time depends on circumstances such as the availability of help, and the speed with which a task needs to be done.

Household energy requirements across the household developmental cycle are modelled for a rural Gambian population, assuming different numbers of offspring and ages at which children cease to be an economic burden on the household by engaging in subsistence work. The peak energy requirement of a household with eight children is more than twice that of a household with two children, while the EDR of the household with eight children is more than three times that of the household with two children. Households with more than two offspring are predicted to spend large portions of their lifecycle with their energy requirements being in excess of the ability to produce. One way in which the energetic burden could be reduced is by children contributing to the household economy. If children become economically productive to the extent of being able to provide the equivalent of their energy needs by the age of 10 years, then households can support up to eight children without energy requirements exceeding the ability to produce food. This is the traditional strategy in rural Gambia, where children begin to assist in agricultural work from about the age of 10 years. Therefore, child labour can be viewed as a strategy whereby large numbers of offspring can be supported, thus maximising reproductive success.

8   *Energetics and human evolution*

The nature and distribution of resources in nature are prime conditioning factors in adaptation through evolutionary time (Kaplan & Hill, 1992). Adaptive features are accounted for in terms of fitness costs and benefits, on the assumption that they should be selected for when the cost/benefit ratio of these features are more optimal in the context of reproductive success than alternative features.

Compared with other primate species, humans and hominids have and had larger brain and body sizes, slower rates of maturation, differing foraging behaviours (Leonard & Robertson, 1992, 1994), and locomotive patterns (Fleagle, 1992), while possessing an optimum body temperature which is related to heat and water balance (McArthur & Clark, 1987). For any species, it is advantageous to use as little energy as possible in the processes of growth and development, bodily maintenance and physical activity related to subsistence and reproduction; balancing the energy needs of individuals and groups involves the ability to obtain an adequate diet and having a body design that minimises energetic and nutritional costs.

Various authors have modelled the energy metabolic consequences of hominid encephalisation (Foley & Lee, 1991; Leonard & Robertson, 1992, 1994), body size (Leonard & Robertson, 1992, 1994; Wheeler, 1993), body shape (Wheeler, 1993) and bipedalism (Wheeler, 1991; Foley, 1992). In this chapter, several of these models are described and compared.

It has been proposed that the ability of humans to run long distances in the heat was a selective advantage in that it allowed them to chase down beasts of medium to large body size by running them to exhaustion (Bortz, 1985). This is easy to envisage in hunter–gatherer societies, since there are descriptions of the running-down of springbok, eland, kudu or hartebeest until they can go no further by Kalahari bushmen (Dorman, 1925; Silberbauer, 1981), and of antelope by the !Kung (Campbell, 1974). In the second part of this chapter, the locomotive and energetic costs of running long are related to the possible benefits of this ability.

166

Bipedalism

In general, foraging efficiency can be increased by: (1) consuming new food items that give higher energy returns; (2) increased use of existing resources; or (3) both of these. Meat and other animal products provide far greater dietary energy density than do foraged foods such as leaves or fruits. Small increases in the amount of dietary energy made available through meat eating could have had major evolutionary consequences, by allowing improved nutritional state, improved fecundity and larger brain and body size. Meat eating may also represent an expansion of the resource breadth beyond that of non-human primates, while technology and cooperative foraging may have increased foraging returns by reducing energetic costs (Foley, 1992). In this way, the energy costs of encephalisation could have been subsidised.

Bipedalism, a characteristic of all known hominids, may have helped to reduce the energetic costs of foraging. Foley (1992) examined this by modelling the energy expended in travelling a number of distances by a 40 kg chimpanzee and a hypothetical bipedal hominid of the same weight. using speeds and energy costs of locomotion estimated by Rodman and McHenry (1980). At the same body weight, a bipedal hominid would have been able to travel up to 11 km for the same level of energy expenditure as a chimpanzee used over a 4 km distance. Furthermore, at larger body size. bipedal hominids were likely to have been more energy efficient than chimpanzees, to the extent that a 53 kg hominid (about the size of *Homo erectus*) would have been able to travel 14 km, while a 57 kg hominid (about the size of early *Homo sapiens*) would have been able to travel a 13 km distance (Foley, 1992). Thus, the greater energetic efficiency of bipedalism could have allowed an increased foraging range area, or greater body size, or both. However, the benefits of bipedalism could have only occurred if foraging was practised over longer distances and in specific ecological conditions. The energetic efficiency of bipedalism would have been gained at the expense of, for example, being able to move efficiently in a woodland or forested environment, as arboreal primates are able to do. Therefore, the energetic advantages of bipedalism could only apply to conditions such as those that were extant 2 million years ago in the East African savannahs, where larger day ranges could confer a foraging advantage, or larger body size could confer a thermoregulatory advantage.

Slower rate of maturation and larger body size

The hominid adaptation of parental investment, through complete provisioning of offspring and their instruction in survival skills, has had a large selective value over the past 2 million years (Bogin, 1990). Prolonged

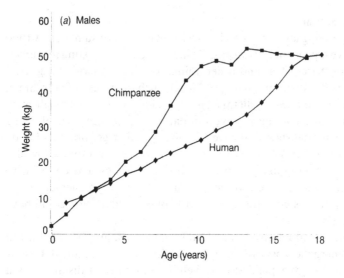

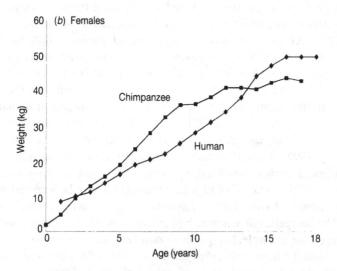

Figure 8.1. Weight by age of chimpanzees and of humans (from a Somali population (Gallo & Mestriner, 1980)); (a) males; (b) females.

somatic growth, relatively advanced central nervous system growth and delayed sexual maturation in humans relative to other species would have promoted these activities. These in turn would have balanced low reproductive output by improved survivability to reproductive age.

In Fig. 8.1, the weight–growth curve of one human population and chimpanzees illustrates fundamental differences in the human growth pattern compared with that of a non-human primate species. The human population chosen is one from Somalia (Gallo & Mestriner, 1980), because adult Somalis have similar adult body weight to chimpanzees and weight by age data are available from early childhood to adulthood. Apart from the first 2 years of life, human growth is slower than that of the chimpanzee. Humans also reach maturity much later, at around 18 years of age, compared with about 10 years of age for male chimpanzees. There is also a much smaller sex difference in body weight in humans, especially in later childhood.

The prolonged growth period of humans is metabolically important, since the energy cost of growth at any stage of development is less than in other mammals (Brody, 1945). Indeed, it has been argued that growth during the early years allows adjustment of body size or proportion to environmental stress, and that slow growth prolongs the time over which body size is relatively small, reducing the total energy requirements for maintenance and growth compared with adult requirements (Bogin, 1988). It therefore reduces the economic burden placed on parents, and minimises any competition for food which might affect parent–offspring or younger–older sibling relations. From an evolutionary viewpoint, it can be argued that the period of extended dependency itself evolved as a sensitive mechanism for enabling increases or decreases in adult body size, whether independently of, or more probably in association with, the intellectual and social characteristics emphasised by Bogin (1990).

There are a number of ways in which increased body size could lead to selective advantage. Foley (1987) has emphasised capacity to exploit broader dietary niches in larger foraging areas, greater mobility and prey size and increased sociality, with altruism enhancing offspring survival. Hill (1982) has suggested several selection pressures that would promote greater predation or specialisation in hunting. The postulated implications of this for reproductive and behavioural patterns centre on raised parental investment in offspring through male provisioning and female child care. However, body size should also be included. According to Hill (1982), these factors could be seen as critical to the emergence of hominids as distinct from other primates. The variable efficiency of hunting would then have been a necessary condition of the later successful expansion of *Homo erectus* and *Homo sapiens* (Gordon, 1987).

The evolution of larger brains

Encephalisation has been shown to correlate with a large number of life history, ecological, and social parameters (Table 8.1). The possession of a

Table 8.1. *Known relationships between relative brain size and life history, ecological and social parameters*

Parameters	Significance in hominid evolution
Life history	
Gestation length (1)	Relatively unchanged
Lifespan (1)	Greatly extended
Neonatal weight (1)	Smaller
Weaning age (1)	Greatly reduced in agricultural and horticultural populations (5)
Age of first reproduction (1)	Slightly later than African apes (5)
Interbirth interval (1)	Similar to gorilla, shorter than chimp (5)
Ecological	
Home range size (2)	Greatly enlarged (6)
Dietary quality (2)	Improved (7)
Social	
Group size (8)	Larger in modern humans, probably larger in most hominids (9)
Social complexity (3)	Increased
Communitication (4)	Language and symbolism

Sources: 1, Harvey, Martin & Clutton-Brock (1987); 2, Clutton-Brock & Harvey (1980); 3, Humphrey (1976); 4, Parker & Gibson (1979); 5, Lee (1989); 6, Foley (1987); 7, Hayden (1981); 8, Dunbar (1992); 9, Foley & Lee (1989).

large brain may be favoured in situations where animals must forage widely, exploit patchy foods or live for a long time in large social groups (Foley, 1992). Such conditions are extant for most large mammals, and the fact that marked encephalisation is found only in humans suggests that there are constraints and that only when these are released is the pattern observed for hominids able to occur.

Brain size is believed to scale to body weight to the power of 0.76 (Martin, 1983); a similar scaling to that seen between basal metabolism and body weight (Brody, 1945). Figure 8.2 shows the encephalisation quotient (EQ) of a number of hominids, calculated using the formula (Martin, 1983):

$$EQ = 0.248 W^{0.76}$$

The EQ has increased across time and is greater in *Homo* than in *Australopithecus*, *Homo sapiens* having by far the largest value. Martin (1989) has argued that brain size is constrained by the metabolic costs of growth and that most species are unable to exceed certain limits, however poorly defined these limits are at present.

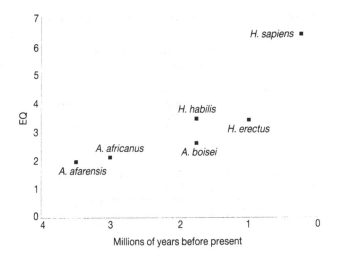

Figure 8.2. Hominid encephalisation quotient (EQ) (Martin, 1983) relative to time (data from Aiello & Dean, 1990; from Foley, 1992).

Foley and Lee (1991) modelled the energy cost of encephalisation and concluded that the difference in total daily energy requirement resulting from differences in brain size between chimpanzees and humans up to the age of 18 months is about 9%. The foraging implications of this are that in some way human mothers must increase the young child's level of nutritional intake compared with the hominoid baseline. Therefore, it might be expected that there should be a close link between expansion of the brain and changes in foraging behaviour and measures of parental effort, regardless of the selective pressures leading to encephalisation. Indeed, Table 8.1 shows that there are significant correlations between greater brain size and larger home range, higher quality diet, larger group size and increased social complexity; all of which arguably lead to greater foraging success, both quantitatively and qualitatively.

A number of foraging and dietary changes have been proposed for hominids during the course of evolution. These include: (1) meat eating; (2) increased extractive efficiency through technology; (3) use of underground resources; (4) food sharing; and (5) central place foraging (Foley & Lee, 1991). If the higher costs of encephalisation are the primary causes of these shifts, then there is no reason to expect these to occur before the appearance and evolution of the genus *Homo*. Evidence of dietary change (Andrews & Martin, 1991) and the appearance of stone tools associated with large mammal bone processing (Blumenschine, 1991) coincident with the origins of *Homo* suggest this to be the case.

Table 8.2. *Measured and estimated resting metabolic rate and metabolic rate of brain of hominids*

	Body weight (kg)	RMR (kJ)	Brain MR (kJ)	RMR – Brain MR (kJ)
Australopithecus afarensis	37	4402	397	4005
A. africanus	35	4243	431	3812
A. boisei	44	5037	494	4543
Homo habilis	48	5339	594	4745
H. erectus	53	5753	837	4916
H. sapiens (early)	57	6075	996	5079
H. sapiens (modern)	54	5975	1319	4656

Source: From Leonard & Robertson (1992).

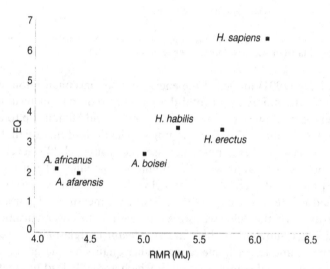

Figure 8.3. Hominid encephalisation quotient (EQ) (Martin, 1983) relative to estimated RMR (Leonard & Robertson, 1992).

In a comparison of body size and the maintenance energy expenditures in hominids, Leonard and Robertson (1992) indicated that not only was body mass greater in *Homo* than in *Australopithecus*, but also that the estimated resting energy expenditure was greater, as was brain weight and the metabolic rate of the brain (Table 8.2). In addition, there appears to be a curvilinear relationship between predicted RMR and EQ, with *Homo sapiens* having the highest values for both (Fig. 8.3). There is a 35% difference in estimated resting energy needs of *Australopithecus* and *Homo sapiens*. Leonard and Robertson (1992) suggest that key differences lie between *Homo habilis*, *Homo erectus* and *Homo sapiens*. That is, total resting energy needs increased substantially between *Homo habilis* and *Homo erectus*, largely as a result of increased body size, while the proportion of dietary energy required by the brain increased most dramatically between *Homo erectus* and *Homo sapiens*.

Man and wildebeest: the energetics of running long

Bipedalism could have allowed larger home ranges and greater body size, while the hunting of animals could have extended the resource breadth of hominids beyond that of non-human primates and made available resources of higher dietary energy density than most other potential foods. Although hominids ate animals of a variety of sizes from mouse-size upwards (Isaac & Crader, 1981), larger mammals such as wildebeest and antelopes may have been commonly taken. Although it is possible that such animals were hunted down using cooperative foraging practices, it has been suggested that in large part they may have been chased down by individuals utilising the ability to run long (Bortz, 1985). This is supported by evidence of contemporary hunter–gatherers running down a number of different middle-to-large size species of mammals (Dorman, 1925; Campbell, 1974; Silberbauer, 1981). How important was the ability to run long likely to have been?

Humans are not the fastest of runners. The maximum speeds of mammals ranging from 20 kg gazelles to 1000 kg giraffes range between 10 and 14 m/s for all species except buffalo, which reach only 7 m/s (Alexander, Langman & Jayes, 1977). Fast human sprinters, however, reach maximum speeds of 11 m/s. The impression given by Alexander (1992) is that humans are slow in comparison with mammals of similar size. However, they are able to accelerate from a stationary position at about 10 m/s² (Ballreich & Kuhlow, 1986), about twice the initial acceleration of gazelles and wildebeest (Alexander & Maloiy, 1989). Therefore, it would seem plausible that a gazelle could be captured without running down, provided that the hunter had the advantage of surprise, while the relative advantage of the human in acceleration and maximum speed over the wildebeest suggests that running to ground need not have been necessary.

Sprinting speeds require anaerobic as well as aerobic metabolism and cannot be sustained for prolonged periods. Maximum speeds that can be sustained by aerobic metabolism only are much lower. Maximum aerobic speeds of wild mammals between 10 and 25 kg mass ranged from 2 to 6 m/s, while men running on treadmills can sustain speeds of up to 6 m/s, purely by aerobic processes. Since the maximum aerobic speed of humans is very

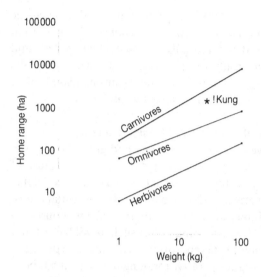

Figure 8.4. Graph on logarithmic coordinates of home range against body mass for carnivores, omnivores and herbivores, with one representative human population, the !Kung bushmen of the Kalahari desert (modified from Alexander, 1992).

close to that predicted for their body size (Garland, Geiser & Baudinette, 1988), it is plausible that humans, and possibly hominids, were able to run to ground a number of species of mammals, including some that were of the same size, even though it may not have been necessary to run wildebeest to ground.

The possibility of running animals to ground is likely to have come with larger body size, in association with increased home range. This is because the cost of aerobic transport declines per unit of body mass as body mass increases. Figure 8.4 shows the home range against body mass for a number of species. Notably, carnivores have larger home ranges for a given body mass than omnivores or herbivores. In energetic terms, this means that larger hunting species have the energetic economies of larger body size, which allow them to forage more widely, and the ability to successfully chase down other species smaller than themselves. Hominids had the added thermoregulatory advantages of smaller exposure to sunlight as a result of bipedalism and the ability to lose heat by evaporation because of their naked skin (Wheeler, 1992).

It would seem that once the spiral of success had started, it was unbeatable in competition with other species; larger body size, bipedalism and greater brain size conferred advantages in the ability to utilise resources. The upper limits on success are likely to have arisen from within-species competition. Hominids are likely to have competed with each other, either at individual or band level, for high-quality resources near the top of the food chain. Out-migration would have helped reduce competition, which would have arisen from the comparative reproductive success of hominids over other species. The primary stress is likely to have been the availability of dietary energy, regardless of source and regardless of whether that energy came from protein, carbohydrate or fat.

Summary

Compared with other primate species, humans and the hominids from which they evolved have and had larger brain and body sizes, slower rates of maturation, differing foraging behaviours and locomotive patterns. The energy metabolic consequences of hominid encephalisation, body size, body shape and bipedalism have been modelled by various authors. Furthermore, the ability to run long distances has been argued to be of selective advantage in human evolution since this could allow medium to large animals to be run to exhaustion. In this chapter, the energetics of encephalisation and bipedalism are considered, and the possible selective advantages of greater body size and of running long are examined.

Bipedalism is likely to have been energetically more efficient than quadrupedalism and could have allowed hominids to have increased foraging ranges, or greater body size, or both. However, the energetic advantage of bipedalism could only apply to conditions such as those that were extant 2 million years ago in the East African savannahs, where larger day ranges could confer a foraging advantage or larger body size could confer a thermoregulatory advantage.

In a comparison of body size and the maintenance energy expenditures in hominids, Leonard and Robertson (1992) suggested that not only was body mass greater in *Homo* than in *Australopithecus* but also that the estimated resting energy expenditure was greater, as was brain weight and the metabolic rate of the brain. Total resting energy needs increased substantially between *Homo habilis* and *Homo erectus*, largely as a result of increased body size, while the proportion of dietary energy required by the brain increased most dramatically between *Homo erectus* and *Homo sapiens*.

Humans are slow runners in comparison with mammals of similar size. However, as large mammals, hominids had the energetic economies of larger body size which allowed them to forage more widely and gave them the ability to successfully chase down other species smaller than themselves. They also had the thermoregulatory advantages of smaller exposure to sunlight as a result of bipedalism and the ability to lose heat by evaporation because of their naked skin.

9 Energy balance and seasonality

The appreciation that seasonality of climate and subsistence productivity can lead to seasonal variation in energy balance (Fox, 1953; Haswell, 1953, 1981; Hauck, Thorangkul & Rajatasilpin, 1960; Annegers, 1973; Schofield, 1974; Bayliss-Smith, 1981; Chowdhury, Huffman & Chen, 1981; Longhurst & Payne, 1981) has resulted in considerable attention being paid to the recording of such variation (Huss-Ashmore & Goodman, 1988; Ferro-Luzzi, 1990a; Ferro-Luzzi et al., 1990; Schultinck et al., 1990; Ulijaszek et al., 1994) and the implications this might have for different aspects of human biology including reproductive performance (Prentice, 1980; Panter-Brick et al., 1993; Ulijaszek, 1993a), child growth (Bogin, 1979; Billewicz & McGregor, 1982; Rosetta, 1988b; Cole, 1993), nutritional status (Benefice et al., 1984; Loutan & Lamotte, 1984; Becker et al., 1986b; Sepulveda, Willett & Munoz, 1988; Lawrence et al., 1989; Payne, 1989; Ferro-Luzzi, 1990b; Simondon et al., 1993) and within-household food allocation (Abdullah & Wheeler, 1985; Wheeler & Abdullah, 1988).

Seasonality of energy balance has been examined in relation to possible physiological changes that might occur in response to changing energy balance (Durnin, Drummond & Satyanarayana, 1990; Ferro-Luzzi et al., 1990; Schultinck et al., 1990; Shetty & Kurpad, 1990); and behavioural strategies which might be effective in ameliorating them (Watts, 1986, 1988; Messer, 1988; Thomas & Leatherman, 1990; Huss-Ashmore, 1993). Models have also been elaborated in which patterns of seasonal energy imbalance have been related to different strategies of food-use (Dugdale & Payne, 1986), and the possible impact of seasonal weight loss has been related to physical work capacity and productivity (Ulijaszek & Strickland, 1993a). In this chapter, the relationships between subsistence performance, energy imbalance and reproductive performance are examined, using information available for rural Gambian agriculturalists. Subsequently, evidence for physiological adaptation to seasonal change in adult body weight is reviewed.

Seasonality, subsistence and reproductive performance
Figure 9.1 gives output–input ratios, or the ratio of energy expended in food production or the food quest and the energy obtained in that quest,

176

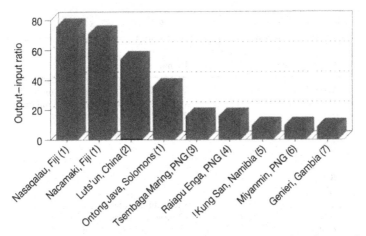

Figure 9.1. Output–input ratios for selected populations. Sources: 1, Bayliss-Smith, 1977; 2, Fei & Chang, 1945; 3, Rappaport, 1968; 4, Waddell, 1972; 5, Lee, 1969; 6, Morren, 1977; 7, Haswell, 1953.

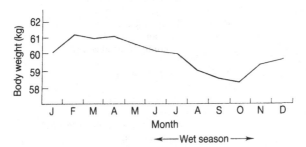

Figure 9.2. Seasonal variation in body weight, rural Gambian males (from Fox, 1953).

for a number of traditional societies. A value above about 8 is suggestive of the ability to generate surpluses, which could be used to buffer against possible seasonal unpredictability. The value for the Gambia, as estimated by Haswell (1981) is 8.6, suggesting that rural Gambian populations may be little able to buffer themselves against seasonal energetic stress. Figure 9.2 shows the seasonal weight change that occurs in adult male Gambians.

As in most tropical, agriculture-based societies, the subsistence season is governed by the annual rains, which in the Gambia fall between June and October. The early part of the wet season is a time in which food shortages are experienced, physical work output is required to clear and plant the fields and the transmission of malaria and diarrhoeal diseases is high. During the months September to November, the rains decline then cease.

Table 9.1. *Reported post-harvest losses in staples*

	Country/region	Weight loss (% of crop)	Remarks
Rice	West Africa	6–14	Weight loss (%): drying, 1–2; on-farm storage, 2–10; parboiling, 1–2; milling, 2–10
	Sierra Leone	10	
	Uganda	11	
	Rwanda	9	
	Sudan	17	Central storage
Maize	Benin	8–9	
	Ghana	7–17	
		15	8 months' storage
	Ivory Coast	5–10	12 months' storage on cob
	Kenya	10–23	4–6 months' central storage
		12	Hybrid maize, hotter regions, 6 months
	Malawi	6–14	Weight loss (%): drying, 6; on-farm storage, 8
		min. 10	Hybrid
	Nigeria	1–5	On-farm storage
		5.5–70	6 months' on-farm storage
	Rwanda	10–20	On-farm storage
	Tanzania	20–100	Unspecified storage
		9, 14, 67	3, 6, and 9 months
	Togo	5–10	6 months' central storage
	Uganda	4–17	
	Zambia	9–21	On-farm storage
Wheat	Sudan	6–19	
	Zimbabwe	10	On-farm storage
Barley	Sudan	17	
Millet	Mali	2–15	Weight loss (%): on-farm storage, 2–4; central storage, 10–14
	Nigeria	0.1–0.2	On-farm storage
	Sudan	14	Central storage
	Zambia	10	On-farm storage
	Zimbabwe	10–15	On-farm storage
Sorghum	Nigeria	0–37	On-farm over 26 months
	Sudan	6–20	Central storage
	Zambia	0–10	Local varieties, negligible; high-yielding varieties, 10% weight loss
	Zimbabwe	25	On-farm storage
Legumes	Ghana	7–45	Shelled beans, 1–5 months; unshelled beans, 22
	Nigeria	5.4	Cowpeas
		1–2	Cowpeas stored 3 months in shell
		4.5	Ground-nuts
	Kenya	30	On-farm storage
	Sudan	4–27	Ground-nuts, central storage

Table 9.1. (*cont.*)

	Country/region	Weight loss (% of crop)	Remarks
	Swaziland	5	Ground-nuts, insects and mould
	Uganda	9–18.5	Ground-nuts, mainly insects and mould
	Zambia	40	Cowpeas
		5	On-farm storage, ground-nuts
Roots/tubers	Ghana	10–20	
	Nigeria	15–60	
		10–50	Yams
	Rwanda	5–40	

Source: From National Academy of Sciences (1978).

Transmission of infection remains high, work output, particularly in the harvesting period, is high, although food becomes increasingly available with the first harvest. Following this, the period December to May involves little physiological stress. Weight loss occurs in the period May to October as a result of the energy imbalance that arises from multiple, coincidental stresses during the west season (Fig. 9.2).

Payne and Dugdale (1977) developed a model that simulates the regulation of energy balance in adults. Using this, they examined the implications for long-term energy balance of different food consumption patterns against possible storage losses of food harvested, but not consumed, for Gambian farmers (Dugdale & Payne, 1986; Payne, 1989). In common with many other agricultural communities in the developing world, Gambian farmers do not consume the largest amounts of food at the time when work output is the heaviest on land preparation and harvesting. Rather, they eat most during the immediate post-harvest period, when energy expenditure is lowest (Fox, 1953).

Dugdale and Payne (1986) compared four different strategies of food allocation: (1) the pattern of energy intake observed by Fox (1953); (2) an exaggerated post-harvest feasting pattern, falling off rapidly to a much reduced level; (3) constant daily intake across the entire year; and (4) a pattern of food intake which matches expenditure, maintaining constant zero energy balance. These were set against differing levels of post-harvest storage losses, which can vary enormously (Table 9.1). The most successful strategy in this modelling procedure was the one which gave the highest body weight across all seasons; that is, the greatest amount of food harvested went to sustain the body mass of the farmers. Table 9.2 gives the results of this simulation.

As expected, overall body weight and weight at the beginning of the

Table 9.2. *Simulation of seasonal changes in body weight (kg) with different strategies of food use and different rates of storage loss, male Gambian farmers*

Theoretical annual storage loss %	Body weight (kg) for different pattern of food use			
	Observed	High post-harvest feasting	Constant intake	Intake to maintain constant weight
Average year round body weight				
0	61.5	60.8	62.2	65.0
10	56.8	57.7	55.9	55.4
30	49.1	49.9	48.3	47.2
Maximum body weight change across the year				
0	6.5	7.7	5.2	0
10	6.5	7.6	5.2	0
30	6.4	7.4	5.3	0
Body weight at the start of the heavy work period				
0	64.2	65.1	65.1	65.0
10	59.4	61.9	58.9	55.4
30	51.8	54.1	51.3	47.2

Source: From Dugdale & Payne (1986).

heavy work period is lower the greater the post-harvest storage loss, regardless of food consumption pattern. However, comparing the different food-consumption patterns is quite revealing. If there were no storage losses, the best strategy would be for food consumption patterns to match energy expenditure, since this would ensure maximum body weight across the year. However, with annual storage losses of 10 and 30%, both patterns of high post-harvest consumption lead to higher overall body weight than do constant intake, or constant weight patterns of consumption. Furthermore, with storage losses, the body weight at the start of the heavy work season is greater for the two patterns of high post-harvest consumption than it is for patterns of constant weight and constant intake. Therefore, the high post-harvest intake observed in Gambian farmers can be rationalised as defending the highest body weight, both overall and at the beginning of the wet season when the ability to perform hard work is most important. Some of the ways in which physical work capacity and performance can be influenced by body size and nutritional state are discussed in Chapter 6.

For Gambian women, energy intake, assessed from the differences between energy expenditure and energy balance (determined from body

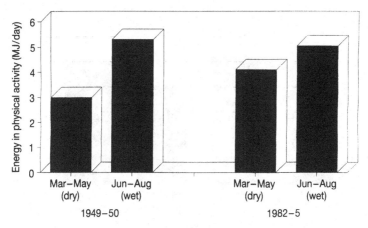

Figure 9.3.   Estimated energy expended in physical activity by rural Gambian women, 1949–50 and 1982–5 (from Haswell, 1953; Lawrence & Whitehead, 1988).

weight changes), is greatest between November and January, lowest between August and October. The latter period is the one of greatest negative energy balance and is more closely related to low energy intake than to seasonally higher energy expenditure (Ulijaszek, 1993b). This is one factor among several influencing reproductive performance in this group.

Although the survival to reproductive age of as many children as possible is the most direct measure of reproductive performance, other, proximate factors can provide a wealth of information about the fine ecological tuning that occurs with this system. Proximate factors include: (1) measures of fertility; (2) pregnancy outcome (one expression of which is birthweight); (3) lactational performance; and (4) child mortality. These are linked to seasonality in the rural Gambian population and, for analytical purposes, have been reduced to the following: (1) food availability and intake of dietary energy; (2) energy expenditure (and also how these first two factors affect energy balance); and (3) infectious disease, as it influences pregnancy outcome and child survivorship. The relative importance of seasonal variation in these factors for reproductive performance has been examined for Gambian women (Lawrence *et al.*, 1989; Ulijaszek, 1993b) and is summarised here.

A comparison of seasonality in energy expenditure in physical activity of women in 1949–50, and 1982–5 is given in Fig. 9.3. It appears that there has been a decline in seasonal variation in activity levels, mainly resulting from the greater dry season energy expenditure in 1982–5 compared with

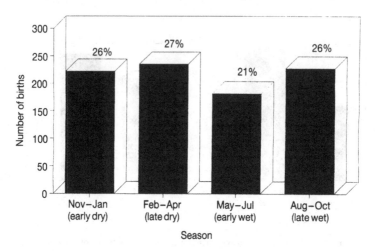

Figure 9.4.   Seasonality of birth rate in rural Gambia (from Billewicz & McGregor, 1981). Values above graph are the percentage of total births in the year.

1949–50. There is little difference in wet season activity levels between these two periods of measurement. The difference in dry season energy expenditure in activity is in the order of 1 MJ per day and can be attributed largely to increases in the time spent performing household maintenance activities, including food preparation, drawing and carrying water, washing clothes and dishes, and so on. In 1949–50, women would spend an average of 3.5 hours each day in such tasks (Haswell, 1953), while in 1982–5, the value reported was 4.8 hours per day (Lawrence & Whitehead, 1988). Time spent in agricultural activities showed little difference across time, for either time of year.

Billewicz and McGregor (1981) demonstrated clear seasonality of birth rates, the lowest number of births taking place at the beginning of the wet season (Fig. 9.4), this effect being attributed to the cumulative effects of heavy work load and negative energy balance across the wet season (Ulijaszek, 1993b). Total daily energy expenditure for pregnant and lactating women in the early wet season (May to July) is higher than at any other time of year, including the late wet season, August to October (Lawrence & Whitehead, 1988). Energy balance (estimated from changes in body weight and skinfold thicknesses) is positive in the dry season, negative in the wet (Fig. 9.5).

There is some seasonal variation in 12-hour breastmilk output, lowest in the August to October period. With respect to protection from ovulation

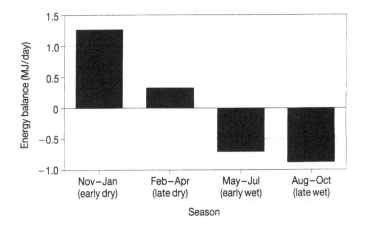

Figure 9.5. Energy balance (estimated from body weight and skinfold thickness measures) of rural Gambian women by season (from Lawrence *et al.*, 1989).

by mechanisms associated with lactation, it is unlikely that these are particularly important during the wet season, if Prentice *et al.* (1986a) are to be believed. They suggest that there is a strong drive towards milk synthesis in Gambian women, milk output being limited, not by food intake, but controlled by the characteristics of the mother–infant pair. This view contradicts evidence from Bangladesh, where the amount of time spent suckling during the times of high work output and throughout the wet season is higher than during the dry season and when work output is low (Chowdhury *et al.*, 1981). However, Lunn (1994) has argued that the post-partum amenorrhoea seen in Gambian women is more likely to be a function of energetic stress associated with lactation, than of lactation *per se* (see Chapter 5). Following from this, the influence of breastfeeding on wet season fertility may be to potentiate the ovulatory-inhibiting effects of negative energy balance.

Pregnant Gambian women show wet season–dry season differences in energy balance, at all stages of pregnancy (Table 9.3), one outcome of which is a difference in mean birthweights between the wet and the dry season (Prentice *et al.*, 1981). Wet-season births are about 200–300 g lighter than are dry-season births, and this may be related to the higher early infancy mortality seen during the wet season.

Table 9.4 gives mortality rates by season of birth and age for rural Gambian children. Those children born between August and October and between November and January have the lowest birthweights and the

Table 9.3. *Estimates of energy balance, from changes in body composition for pregnant Gambian women*

Months of pregnancy	Energy balance	
	Dry season (MJ/day)	Wet season (MJ/day)
0–2.9	+1.16	−0.31
3–5.9	+0.61	−0.22
6–9	+0.84	+0.10

Source: From Prentice *et al.* (1981).

Table 9.4. *Young-child mortality in Keneba, the Gambia*

Season of birth	0–3 month mortality rate (per 1000 live births)	Chances of dying by age	
		1 year (%)	5 years (%)
Dry season			
Nov.–Jan.	135	32	54
Feb.–Apr.	85	21	49
Wet season			
May–July	88	15	40
Aug.–Oct.	123	25	55

Source: From Billewicz & McGregor (1981).

highest mortality rates in the first 3 months of life. Although 3-monthly mortality rates are highest for all birth cohorts when they are in the wet season (Billewicz & McGregor, 1981), the November to January birth cohort has the highest overall mortality rate in the first year of life, while the May to July birth cohort has the lowest mortality rate for the same period. The latter cohort also has the greatest proportion of children surviving to the age of 5 years. Children born in May to July enter the wet season with higher average birthweights than children born later on in the wet season. Breastfeeding is likely to shield them from excessive exposure to diarrhoeal diseases for most of the wet season; the introduction of weaning foods taking place when the worst of the wet season is over. Most of the infant mortality in rural Gambia is caused by the interactive effects of undernutrition and infection; this synergism is greatest in the wet season, but does still operate at a lower level in the dry season when the incidences of respiratory

Table 9.5. *Proportion of all days spent in the fields by rural Gambian women, by season*

Season	28 weeks pregnant–4 weeks post-partum (%)	Early pregnancy and later lactation (%)
Dry	30	26
Early wet	67	81
Late wet	50	77
Harvest	31	42

Source: From Lawrence & Whitehead (1988).

tract infection and diarrhoea are lower if not completely absent (Rowland *et al.*, 1981).

Comparing the mortality of children by season of birth (Table 9.4) with seasonality of birth rates (Fig. 9.4), it is apparent that the birth cohort with greatest young-child survivorship, May to July, is the one with the lowest proportion of births compared with other times of the year. It would appear, therefore, that women who conceive late in the wet season have an advantage in reproductive performance compared with women who conceive at other times of the year. There must be constraints on conception at this time, since it seems logical that all women might seek to maximise their reproductive performance in this way.

The obvious constraint, already discussed, is energetic. Another might be the effect that high birth rates at the beginning of the wet season might have on work schedules. It has been shown that, in the wet season, women in late pregnancy and early lactation go to the fields to work for fewer days than do women who are in early pregnancy or late lactation (Lawrence & Whitehead, 1988) (Table 9.5). When the proportion of all days spent in the fields is crudely translated into days per 7-day week, women in late pregnancy and early lactation spend 4.7 and 3.5 days in the fields during the early wet and late wet seasons, respectively, compared with 5.7 and 5.4 days for women in early pregnancy and for women in late lactation. Furthermore, on the days when women in late pregnancy and early lactation go to the fields, they expend less energy at work than their counterparts who are not at either stage of the reproductive cycle. If the energy costs of pregnancy are in large part borne by this type of reduction in physical activity, there may be limits to the amount of accommodation possible, particularly when set in the context of a subsistence system that does not generate much in the way of surplus.

Table 9.6. *Mean body weight changes in different rural populations in developing contries*

Location	Body weight (kg)		Reference
	Men	Women	
Cameroon, Massa	4.4–5.9	4.1	de Garine & Koppert, 1988
Cameroon, Mussey	4.1–4.2	3.5–4.0	de Garine & Koppert, 1988
Burkina Faso	3.7	2.0	Ancey, 1974
Senegal	3.6		Gessain, 1978
Mali, Fulani	3.2	2.1	Hilderbrand, 1985
Niger, Wodaabe	3.1	2.4	Louton & Lamotte, 1984
Gambia	2.9		Fox, 1953
Burkina Faso	2.8	0.7	Bleiberg et al., 1980
Senegal, Ferlo	2.7	2.1	Benefice et al., 1984
Mali, Rimaibe	2.6	2.0	Hilderbrand, 1985
Gambia	2.5	2.5	Billewicz & McGregor, 1982
Miyanmar	2.1		Tin-May-Than & Ba-Aye, 1985
Peru	1.8	1.2	Leonard & Thomas, 1989
Senegal, Serere	1.7–1.8	1.0–1.2	Rosetta, 1986
Niger	1.6		Reardon, 1989
Botswana	1.1–1.7	0.5–2.2	Wilmsen, 1978
PNG, Girawa	1.4	2.0	Spencer & Heywood, 1983
Zaire, Ntommba	0.5–2.2	0.6–1.4	Pagezy, 1982
Zaire, Walese	1.1	1.0	Bailey & Peacock, 1988
Bangladesh	1.1	0.6	Abdullah & Wheeler, 1985
India, Tamil Nadu	0.5	0.1	McNeill et al., 1988
PNG, Ningerum	0.5	0.1	Ulijaszek, 1992b
Gambia		4.9	Lawrence et al., 1989
Lesotho		4.0	Huss-Ashmore & Goodman, 1988
PNG, Tari		3.0	Crittenden & Baines, 1986
Bangladesh		1.9	Chen et al., 1979
Gambia		1.9	Prentice et al., 1981
Ethiopia, Sidamo		1.6	Ferro-Luzzi et al., 1990
Benin, Mono		1.0	Schultinck et al., 1990
Kenya		0.8	Coghill, 1987
India		0.5	Norgan et al., 1989

In the early wet season, women in late pregnancy or early lactation form about 20% of the total work force. Cooperative organisation of work may allow these women, one fifth of the female work force, to spend 1 day per week less in the fields in the early wet season, if the other 80% of women work 1 day more per month. If a greater proportion of women were to give birth at the beginning of the wet season, it might not be possible to buffer them and their babies against energetic stress without a reduction in overall agricultural work output. It is possible that a reduction in work output could lead to a decline in food production that could affect the nutritional well-being of the entire community. Thus, the reproductive advantage of the few would be bought at the cost of energetic and nutritional disadvantage to the entire community (Ulijaszek, 1993b).

Seasonality of energy balance

In most communities where climatic seasonality might affect human biology in some stressful way, economic or socio-cultural avoidance strategies are employed. However, where avoidance is not possible, physiological and behavioural strategies may be brought into play. Among the most common biological phenomena observed is seasonality of weight change. Ferro-Luzzi and Branca (1993) have summarised seasonal weight losses in a number of third-world populations and concluded that loss rarely exceeds 5% of the seasonal high in adults, women having smaller losses than men. Table 9.6 shows mean weight changes in adults in a number of rural third-world communities.

Mean weight loss in most communities is characterised by large within- and between-population variation, and there is evidence that the amount of weight lost in any group is inversely related to weight at the onset of negative energy balance. In Zaire, Pagezy (1982) found that the heavier Oto women lost more weight (1.4 kg) than the lighter Twa women (0.6 kg), while Schultinck *et al.* (1990) and Ferro-Luzzi *et al.* (1990) showed that women with the highest BMIs were the ones which showed the greatest seasonal weight loss, in Benin and Ethiopia, respectively. Summarising data for populations in 30 countries, Ferro-Luzzi and Branca (1993) concluded that third-world farmers show a consistent tendency to keep the depletion of their lean body mass to within 2%, and that people with a smaller BMI tend to lose less weight in absolute terms. This suggests a resistance to mobilise body energy stores when they are low, to safeguard against the depletion of lean tissue.

Patterns of weight loss do not necessarily tie in with areas of highest climatic seasonality, nor is there a consistent pattern of seasonal weight change across all subsistence types (Ulijaszek & Strickland, 1993a). However, in most places, seasonal weight loss in women is smaller than in men. The ecological importance of this might be related to the greater importance of women to the reproductive process, relative to productive performance. For males, if most subsistence tasks are performed within their productive capacity, a larger body size would raise the cost of their maintenance without providing gain. For women however, successful reproduction is the more important concern. On this basis, it can be argued that reduced body mass may carry less functional significance in males than females, weight loss in females carrying with it the possibility of reduced reproductive potential, through anovulation (Lunn, 1994).

Table 9.7. *Accounting of seasonal negative energy balance in three populations of adult women*

	India	Benin	Ethiopia
BMI (kg/m²)	18	21	19
Body weight (kg)	40	50	45
Body weight change (harvest–lean season) (kg)	−0.3	−1.0	−1.6
Energy equivalence of weight change (MJ)	8.8	29.3	46.9
BMR (harvest–lean season) (MJ/day)	3.67–3.50	5.59–5.60	5.73–5.34
BMR change (%)	−5	0	−7
Energy saved by BMR change (harvest–lean season) (MJ)	8.1	0	34.7
Total energy saved (MJ)	16.9	29.3	81.5
Mean energy intake (harvest–lean season) (MJ/day)	8.50–7.91	6.95–6.52	9.25–8.49
Potential energy debt (MJ)	26.4	25.9	68.6
Unaccounted energy debt (kJ/day)	−105	0	+71

Source: Ferro-Luzzi (1990b).

Weight loss can subsidise seasonal energy-intake deficits to a substantial degree. For example, a weight loss of 4 kg, involving predominantly fat, would make 117 MJ available for metabolic processes. This represents less than 1% of the total body energy, and if this loss took place over a 2-month period it would subsidise a negative balance between energy intake and expenditure of about 2 MJ/day. However, under conditions of negative energy balance, the composition of weight loss comprises both adipose and lean tissue, the proportion of lean tissue lost being inversely proportional to the initial size of the energy stores (Forbes, 1987). Weight loss for a person with a BMI of 40 would include only 20% of lean tissue but this rises steeply to 50% for an individual with a BMI of 21, and 70% for a person with a BMI of less than 17 (Ferro-Luzzi & Branca, 1993). Therefore, it is possible that even modest weight losses might have important health and functional consequences (Ferro-Luzzi & Branca, 1993), particularly in individuals with habitually low BMIs.

Energy balance and nutritional status is a dynamic condition and it could be that small reductions in body weight are sufficient to induce metabolic responses, while large changes may simultaneously affect work capacity and performance. With respect to accounting for energy debt in seasonal environments, Ferro-Luzzi (1990b) cites studies where weight loss can: (1) completely account for the energy debt; (2) completely account for the energy debt only in association with reduced BMR; and (3) only play a minor part in coping with negative energy balance (Table 9.7).

In some situations, where initial body fat stores are adequate, loss of body weight might be a good way of adapting to seasonal energy imbalance. Where body weight and fatness are lower, weight loss may sometimes induce a metabolic response, resulting in down-regulation of BMR. However, absolute change in body weight seasonally may be less important than the relative seasonal decline in BMI, which might have consequences for work productivity and fertility.

Seasonal imbalances between intake and expenditure may be small, but important. For example, between the end of April and the end of September, male Gambian farmers lose, on average, 2.4 kg in body weight (Fox, 1953). This is equivalent to a mean weight loss of 16 g/day, which at a lean to fat mass change ratio of 1, gives a negative energy balance of 435 kJ/day. This amounts to about 5% of daily energy intake but could be saved if 20 minutes/day less were spent at performing hard physical work, at a physical activity ratio of 6. This is 5% of a 6-hour working day. That such a small accommodation to negative energy balance is not made by reducing physical activity suggests that they are unable to do this, presumably because this might threaten overall agricultural productivity.

Across a number of studies, weight loss, even if small, is the only universally observed response to seasonal energy imbalance. Changes in BMR appear to follow intake changes more closely than do weight changes, often declining simultaneously with, or even preceeding, weight loss. This ties in with what is known about thyroid hormone responses to underfeeding, and their effects on basal metabolism (Dauncey, 1990). Reduced physical activity, potentially the most powerful mechanism of saving energy, is usually maintained at the expense of weight loss, which refutes a suggestion by Ferro-Luzzi (1990b) that behavioural modifications might minimise the extent of weight loss observed in third-world communities. Reduction in the performance of economically important activities is likely to occur as a last resort, after all other responses have failed to accommodate the energy imbalance.

Summary

The relationships between seasonal variation in energy balance and reproductive performance, child growth and nutritional status are well documented. Such variation has been examined by various authors, in relation to possible physiological changes that might occur in response to changing energy balance, and behavioural strategies that might be effective in ameliorating them. In this chapter, the relationships between subsistence performance, energy imbalance and reproductive performance are examined, using information available for rural Gambian agriculturalists. Subsequently, evidence for physiological adaptation to seasonal change in adult body weight is reviewed.

References

Abdullah, M. (1989). The effect of seasonality on intrahousehold food distribution and nutrition in Bangladesh. In *Seasonal Variability in Third World Agriculture. The Consequences for Food Security*, ed. D. E. Sahn, pp. 57–65. Baltimore: Johns Hopkins University Press.

Abdullah, M. & Wheeler E. F. (1985). Seasonal variations, and the intra-household distribution of food in a Bangladeshi village. *American Journal of Clinical Nutrition*, **41**, 1305–13.

Abrams, B. (1991). Maternal undernutrition and reproductive performance. In *Infant and Child Nutrition Worldwide: Issues and Perspectives*. ed. F. Falkner, pp. 31–60. Ann Arbor: CRC Press.

Acheson, K. J., Campbell, I. T., Edholm, O. G., Miller, D. S. & Stock, M. J. (1980). The measurement of daily energy expenditure – an evaluation of some techniques. *American Journal of Clinical Nutrition*, **33**, 1155–64.

Adair, L. S., Pollitt, E. & Mueller, W. H. (1983). Maternal anthropometric changes during pregnancy and lactation in a rural Taiwanese population. *Human Biology*, **55**, 771–87.

Adams, C. C. (1935). The relation of human ecology to general ecology. *Ecology*, **16**, 316–35.

Adeokun, L. A. (1982). Marital sexual relationships and birth spacing among two Yoruba sub-groups. *Africa*, **52**, 1–14.

Ahlborg, G., Felig, P., Hagenfeldt, L., Hendler, R. & Wahren, J. (1974). Substrate turnover during prolonged exercise in man: splanchnic and leg metabolism of glucose, free fatty acids, and amino acids. *Journal of Clinical Investigation*, **53**, 1080–90.

Aiello, L. C. & Dean, M. C. (1990). *An Introduction to Human Evolutionary Anatomy*. London: Academic Press.

Akerstedt, T., Gillberg, M., Hjemdahl, P., Sigurdson, K., Gustavsson, I., Daleskog, M. & Pollare, T. (1983). Comparison of urinary and plasma catecholamine responses to mental stress. *Acta Physiologica Scandinavica*, **117**, 19–26.

Alam, N., Rahaman, M. M. & Wojtyniak, B. (1989). Anthropometric indicators and risk of death. *American Journal of Clinical Nutrition*, **49**, 884–8.

Alexander, R. McN. (1992). Comparative aspects of human activity. In *Physical Activity and Health*. ed. N. G. Norgan, pp. 7–19. Cambridge: Cambridge University Press.

Alexander, R. McN., Langman, V. A. & Jayes, A. S. (1977). Fast locomotion of some African ungulates. *Journal of Zoology*, **183**, 291–300.

Alexander, R. McN. & Maloiy, G. M. O. (1989). Locomotion of African mammals. *Symposia of the Zoological Society of London*, **61**, 163–80.

Alkire, W. H. (1965). *Lamotrek Atoll and Inter-Island Socio-Economic ties*. Urbana: University of Illinois Press.

Altman, J. C. (1984). Hunter–gatherer subsistence production in Arnhem Land: the original affluence hypothesis re-examined. *Mankind*, **14**, 179–90.

Ancey, G. (1974). Facteurs et Systemes de production dans la société Mossi d'aujourd'hui. *Migration-Travail-Terre et Capital*. Ouagadougou, Haute Volta: Centre ORSTOM.

Anderson, M. A. (1989). The relationship between maternal nutrition and child growth in rural India. PhD dissertation, Tufts University.

Anderson, N. G. (1976). A five year survey of small for dates infants for chromosome abnormalities. *Australian Pediatric Journal*, **12**, 19–23.

Andrews, P. & Martin, L. (1991). Hominoid dietary evolution. *Philosophical Transactions of the Royal Society of London, Series B*, **334**, 199–209.

Annegers, J. F. (1973). Seasonal food shortages in West Africa. *Ecology of Food and Nutrition*, **2**, 251–7.

Arab, L., Whittler, M. & Schettler, G. (1987). *European Food Composition Tables in Translation*. Berlin: Springer Verlag.

Ashworth, A. (1968). An investigation of very low caloric intakes reported in Jamaica. *British Journal of Nutrition*, **22**, 341–55.

Aslan, S., Carruthers, M., Nelson, L. & Lader, L. (1981). Stress and age effects on catecholamines in normal subjects. *Journal of Psychosomatic Research*, **25**, 33–41.

Astrand, P. O. & Rodahl, K. (1986). *Textbook of Work Physiology*, 3rd edn. New York: McGraw-Hill.

Avons, P., Garthwaite, P., Davies, H. L., Murgatroyd, P. R. & James, W. P. T. (1988). Approaches to estimating physical activity in the community: calorimetric validation of actometers and heart rate monitoring. *European Journal of Clinical Nutrition*, **42**, 185–96.

Aw, T. Y. & Jones, D. P. (1989). Nutrient supply and mitochondrial function. *Annual Review of Nutrition*, **9**, 229–51.

Bailey, R. C. & Peacock, N. R. (1988). Efe Pygmies of northeast Zaire: subsistence strategies in the Ituri forest. In *Coping with Uncertainty in Food Supply*. ed. I. de Garine & G. A. Harrison, pp. 88–117. Oxford: University Press.

Bailey, R. C., Head, G., Jenike, M., Owen, B., Rechtman, R. & Zechenter, E. (1989). Hunting and gathering in tropical rainforest: is it possible? *American Anthropologist*, **91**, 59–82.

Bailey, R. C., Jenike, M. R., Ellison, P. T., Bentley, R. R., Harrigan, A. M. & Peacock, N. R. (1992). The ecology of birth seasonality among agriculturalists in central Africa. *Journal of Biosocial Science*, **24**, 393–412.

Bailey, R. C., Jenike, M. R., Ellison, P. T., Bentley, G. R., Harrigan, A. M. & Peacock, N. R. (1993). Seasonality of food production, nutritional status, ovarian function and fertility in Central Africa. In *Tropical Forests, People and Food. Biocultural Interactions and Applications to Development*. ed. C. M. Hladik, A. Hladik, O. F. Linares, H. Pagezy, A. Semple & M. Hadley, pp. 387–402. Paris: UNESCO Publications.

Bairagi, R., Chowdhury, M. K., Kim, Y. J. & Curlin, G. T. (1985). Alternative anthropometric indicators of mortality. *American Journal of Clinical Nutrition*, **42**, 296–306.

Ballreich, R. & Kuhlow, A. (1986). *Biomechanik der Leichtathletik.* Stuttgart: Enke.

Balogh, M., Medalie, J. N., Smith, H. & Groen, J. J. (1968). The development of a dietary questionnaire for an ischaemic heart disease survey. *Israel Journal of Medical Science,* **4,** 195–203.

Bandini, L. G., Schoeller, D. A., Cyr, H. N. & Dietz, W. H. (1990). Validity of reported energy intake in obese and nonobese adolescents. *American Journal of Clinical Nutrition,* **52,** 421–5.

Banerjee, B. & Saha, N. (1981). Energy balance study in pregnant Asian women. *Tropical and Geographical Medicine,* **33,** 215–18.

Banerjee, B., Khew, K. S. & Saha, N. (1971). A comparative study of energy expenditure in some common daily activities of non-pregnant and pregnant Chinese, Malay and Indian women. *Journal of Obstetrics and Gynaecology in the British Commonwealth,* **78,** 113–16.

Banister, E. W. & Brown, S. R. (1968). The relative energy requirements of physical activity. In *Exercise Physiology.* ed. H. B. Falls, pp. 267–322. New York: Academic Press.

Bantje, H. F. W. (1988). Female stress and birth seasonality in Tanzania. *Journal of Biosocial Science,* **20,** 195–202.

Barrell, R. A. E. & Kolley, S. S. M. I. (1982). Cow's milk as a potential vehicle of diarrhoea disease pathogens in a West African village. *Journal of Tropical Pediatrics,* **28,** 48–52.

Barrell, R. A. E. & Rowland, M. G. M. (1979). Infant foods as a potential source of diarrhoeal illness in rural West Africa. *Transactions of the Royal Society of Tropical Medicine and Hygiene,* **73,** 85–90.

Barrows, H. H. (1922). Geography as human ecology. *Annals of the Association of American Geographers,* **13,** 1–14.

Bayliss-Smith, T. P. (1977). Energy use and economic development in Pacific communities. In *Subsistence and Survival: Rural Ecology in the Pacific.* ed. T. P. Bayliss-Smith & R. G. A. Feachem, pp. 317–62. London: Academic Press.

Bayliss-Smith, T. P. (1981). Seasonality and labour in the rural energy balance. In *Seasonal Dimensions to Rural Poverty.* ed. R. Chambers, R. Longhurst & A. Pacey, pp. 30–8. London: Frances Pinter Ltd.

Bayliss-Smith, T. P. (1982a). Ecology of Agricultural Systems. Cambridge: Cambridge University Press.

Bayliss-Smith, T. P. (1982b). Energy use, food production and welfare: perspectives on the 'efficiency' of agricultural systems. In *Energy and Effort.* ed. G. A. Harrison, pp. 283–303. London: Taylor and Francis.

Bayliss-Smith, T. P. & Feachem, R. G. (1977). *Subsistence and Survival: Rural Ecology in the Pacific.* New York: Academic Press.

Becker, S. (1981). Seasonality of fertility in Matlab, Bangladesh. *Journal of Biological Science,* **13,** 97–105.

Becker, S., Chowdhury, A. & Leridon, H. (1986a). Seasonal patterns of reproduction of Matlab, Bangladesh. *Population Studies,* **40,** 457–72.

Becker, S., Black, R. E., Brown, K. H. & Nahar, S. (1986b). Relations between socio-economic status and morbidity, food intake and growth in young children in two villages in Bangladesh. *Ecology of Food and Nutrition,* **18,** 251–64.

Belko, A. Z., Barbieri, T. F. & Wong, E. C. (1986). Effect of energy and protein intake and exercise intensity on the thermic effect of food. *American Journal of Clinical Nutrition*, **43**, 863–9.

Benedict, F. G., Emmes, L. E., Roth, P. & Smith, H. M. (1914). The basal gaseous metabolism of normal men and women. *Journal of Biological Chemistry*, **18**, 139–55.

Benefice, E., Chevassus-Agnes, S. & Barral, H. (1984). Nutritional situation and seasonal variations for pastoralist populations of the Sahel (Senegalese Ferlo). *Ecology of Food and Nutrition*, **14**, 229–47.

Bentley, G. R. (1985). Hunter–gatherer energetics and fertility: a reassessment of the !Kung San. *Human Ecology*, **13**: 79–109.

Bergmann, R. L. & Bergmann, K. E. (1986). Nutrition and growth in infancy. In *Human Growth: a Comprehensive Treatise*. ed. F. Falkner & J. M. Tanner, pp. 389–413. New York: Plenum Press.

von Bertalanffy, L. (1968). *General Systems Theory*. New York: Braziller.

Bertes, N. A. (1988). K'ekchi' horticultural labor exchange: productive and reproductive implications. In *Human Reproductive Behaviour*. ed. L. Betzig, M. Borgerhoff Mulder & P. Turke, pp. 83–96. Cambridge: Cambridge University Press.

Billewicz, W. Z. (1979). The timing of post-partum menstruation and breast feeding: a simple formula. *Journal of Biosocial Science*, **11**, 141–51.

Billewicz, W. Z. & McGregor, I. A. (1981). The demography of two West African (Gambian) villages, 1951–75. *Journal of Biosocial Science*, **13**, 219–40.

Billewicz, W. Z. & McGregor, I. A. (1982). A birth-to-maturity longitudinal study of heights and weights in two West African (Gambian) villages, 1951–1975. *Annals of Human Biology*, **9**, 309–20.

Bingham, S. A. (1991). Assessment of food consumption and nutrient intake. Current intake. In *Design Concepts in Nutritional Epidemiology*. ed. B. M. Margetts & M. Nelson, pp. 154–67. Oxford: Oxford University Press.

Bisdee, J. T., James, W. P. T. & Shaw, M. A. (1989). Changes in energy expenditure during the menstrual cycle. *British Journal of Nutrition*, **61**, 187–99.

Black, A. E., James, W. P. T. & Besser, G. M. (1983). Obesity. A Report of the Royal College of Physicians. *Journal of the Royal College of Physicians of London*, **17**, 5–65.

Black, A. E., Prentice, A. M. & Coward, W. A. (1986). Use of food quotients to predict respiratory quotients for the doubly-labelled water method of measuring energy expenditure. *Human Nutrition: Clinical Nutrition*, **40C**, 381–91.

Black, A. E., Jebb, S. & Bingham, S. (1991). Validation of energy and protein intakes assessed by diet history and by weighed records against total energy expenditure and 24h urinary nitrogen. *Proceedings of the Nutrition Society*, **50**, 108A.

Blackburn, M. W. & Calloway, D. H. (1976a). Basal metabolic rate and work energy expenditure of mature, pregnant women. *Journal of the American Dietetic Association*, **69**, 24–8.

Blackburn, M. W. & Calloway, D. H. (1976b). Energy expenditure and consumption of mature pregnant and lactating women. *Journal of the American Dietetic Association*, **69**, 29–37.

Blaxter, K. (1989). *Energy Metabolism in Animals and Man*. Cambridge: Cambridge University Press.

Bleiberg, F. M., Brun, T. A. & Goihman, S. (1980). Duration of activities and energy expenditure of female farmers in dry and rainy season in Upper-Volta. *British Journal of Nutrition*, **43**, 71–82.

Bleiberg, F. M., Brun, T. A., Goihman, S. & Lippman, D. (1981). Food intake and energy expenditure of male and female farmers from Upper Volta. *British Journal of Nutrition*, **45**, 505–15.

Blume, F. D. (1983). Metabolism and nutrition at altitude. In *Hypoxia, Exercise, and Altitude*. ed. J. R. Sutton, C. S. Houston & N. L. Jones, pp. 311–16. New York: Alan Liss.

Blumenschine, R. J. (1991). Hominid carnivory and foraging strategies, and the socio-economic function of early archaeological sites. *Philosophical Transactions of the Royal Society of London, Series B*, **334**, 211–21.

Blurton Jones, N. & Sibly, R. M. (1978). Testing adaptiveness of culturally determined behaviour: do Bushman women maximize their reproductive success by spacing births widely and foraging seldom? In *Human Behaviour and Adaptation*. ed. N. Blurton Jones & V. Reynolds, pp. 135–57. London: Taylor and Francis.

de Boer, J. O., van Es, A. J. H., Voorrips, L. E., Blokstra, F. & Vogt, J. E. (1988). Energy metabolism and requirements in different ethnic groups. *European Journal of Clinical Nutrition*, **42**, 983–97.

Bogin, B. (1979). Monthly changes in the gain and loss of growth in weight of children living in Guatemala. *American Journal of Physical Anthropology*, **51**, 287–92.

Bogin, B. (1988). *Patterns of Human Growth*. Cambridge: Cambridge University Press.

Bogin, B. (1990). The evolution of human childhood: a unique growth phase and delayed maturity allow for extensive learning and complex culture. *Bioscience*, **40**, 16–25.

Bongaarts, J. (1980). Does malnutrition affect fecundity? A summary of evidence. *Science*, **208**, 564–9.

Bongaarts, J. (1982). The fertility-inhibiting effects of the intermediate fertility variables. *Studies in Family Planning*, **13**, 179–89.

Boothby, W. M. & Sandiford, I. (1922). Summary of the basal metabolism data on 8614 subjects with special reference to the normal standards for the estimation of the basal metabolic rate. *Journal of Biological Chemistry*, **54**, 783–803.

Boothby, W. M. & Sandiford, I. (1929). Normal values for standard metabolism. *American Journal of Physiology*, **90**, 290–1.

Bortz, W. M. (1985). Physical exercise as an evolutionary force. *Journal of Human Evolution*, **14**, 145–55.

Boserup, E. (1986). Shifts in the determinants of fertility in the developing world: environmental, technical, economic and cultural factors. In *The State of Population Theory. Forward from Malthus*. ed. D. Colemen & R. Schofield, pp. 239–55. Oxford: Basil Blackwell.

Boutellier, U., Howald, H., di Prampero, P. E., Giezendammer, D. & Cerretelli, P. (1983). Human muscle adaptations to chronic hypoxia. In *Hypoxia, Exercise, and Altitude*. ed. J. R. Sutton, C. S. Houston & N. L. Jones, pp. 273–81. New York: Alan Liss.

Boutron, M. C., Faivre, J., Milan, C., Gorcerie, B. & Esteve, J. (1989). A comparison of two diet history questionnaires that measure usual food intake. *Nutrition and Cancer*, **12**, 83–91.

Bouverot, P. (1985). Adaptation to altitude-hypoxia in vertebrates. In *Zoophysiology*. ed. B. Heinrich, K. Johansen, H. Langer, G. Neuweiler & D. J. Randall, pp. 120–41. Berlin: Springer-Verlag.

Brenton, B. P. (1988). The seasonality of storage. In *Coping with Seasonal Constraints*. ed. R. Huss-Ashmore, J. J. Curry & R. K. Hitchcock, pp. 45–54. Philadelphia: MASCA Research Papers in Science and Archaeology.

Brody, S. (1945). *Bioenergetics and Growth*. New York: Reinhold.

Brooks, R. M., Latham, M. C. & Crompton, D. W. T. (1979). The relationship of nutrition and health to worker productivity in Kenya. *East African Medical Journal*, **56**, 413–21.

Brotherhood, J. R. (1973). Studies on energy expenditure in the Antarctic. In *Polar Human Biology*. ed. O. J. Edholm & E. K. E. Gunderson, pp. 113–19. London: Heineman.

Brown, K. H., Black, R. E., Robertson, A. D. & Becker, S. (1985). Effects of season and illness on the dietary intake of weanlings during longitudinal studies in rural Bangladesh. *American Journal of Clinical Nutrition*, **41**, 343–55.

Brown, M. L., Worth, R. M. & Shah, N. K. (1968). Diet and nutritional status of the Nepalese people. *American Journal of Clinical Nutrition*, **21**, 875–81.

Brown, T. P. (1987). Work capacity study (1985/6). In *Ok Tedi Health and Nutrition Project 1982–1986. Final Report*. ed. J. A. Lourie, pp. 43–54. Port Moresby: University of Papua New Guinea.

Brun, T. (1992). The assessment of total energy expenditure of female farmers under field conditions. *Journal of Biosocial Science*, **24**, 325–33.

Brun, T. A., Bleiberg, F. M. & Goihman, S. (1981). Energy expenditure of male farmers in dry and rainy seasons in Upper-Volta. *British Journal of Nutrition*, **45**, 67–82.

Brun, T. A., Daxun, J., Geissler, C. et al. (1988). *Food Consumption and Work Capacity in a Tropical Environment in China*. Paris: INSERM.

Brunton, J. A., Bertolo, R. F. P., Winthrop, A. L. & Atkinson, S. A. (1992). Body composition analysis by dual energy X-ray absorptiometry (DEXA) compared to chemical analysis of fat, lean, and bone mass in small piglets. *Proceedings of the International In Vivo Body Composition Conference*, A47.

Brush, G., Harrison, G. A., Baber, F. M. & Zumrawi, F. Y. (1992). Comparative variability and interval correlation in linear growth of Hong Kong and Sudanese infants. *American Journal of Human Biology*, **4**, 291–9.

Buckley, W. (1967). *Sociology and Modern Systems Theory*. Englewood Cliffs, NJ: Prentice Hall.

Bullen, B. A., Skrinar, G. S., Beitins, I. Z., Mering, G. von, Turnbull, B. A. & McArthur, J. W. (1985). Induction of menstrual disorders by strenuous exercise in untrained women. *New England Journal of Medicine*, **312**, 1349–53.

Burnham, P. (1982). Energetics and ecological anthropology: some issues. In *Energy and Effort*. ed. G. A. Harrison, pp. 229–41. London: Taylor and Francis.

Butte, N. F. (1990). Basal metabolism of infants. In *Activity, Energy Expenditure and Energy Requirements of Infants and Children*. ed. B. Schurch & N. S.

Scrimshaw, pp. 117–36. Lausanne: International Dietary Energy Consultancy Group.

Butte, N. F., Wills, C., Smith, E. O. & Garza, C. (1985). Prediction of body density from skinfold measurements in lactating women. *British Journal of Nutrition*, **53**, 485–9.

Caldwell, J. C. & Caldwell, P. (1977). The role of marital sexual abstinence in determining fertility: a study of the Yoruba in Nigeria. *Population Studies*, **31**, 193–213.

Caldwell, J. C. & Caldwell, P. (1981). The function of child-spacing in traditional societies and the direction of change. In *Child-Spacing in Tropical Africa: Traditions and Change*. ed. H. J. Page & R. Lesthaeghe, pp. 73–92. New York: Academic Press.

Campbell, B. B. (1974). *Human Evolution*. Chicago: Aldine de Gruyter.

Campbell, K. L. & Wood, J. W. (1988). Fertility in traditional societies. In *Natural Human Fertility*. ed. P. Diggory, M. Potts & S. Teper, pp. 39–69. Basingstoke: Macmillan Press.

Campbell, P. & Ulijaszek, S. J. (1994). Relationships between anthropometry and retrospective morbidity in poor men in Calcutta, India. *European Journal of Clinical Nutrition*, in Press.

Casimir, M. J. (1991). *Flocks and Food. A Biocultural Approach to the Study of Pastoral Foodways*. Köln: Böhlau Verlag.

Cavalli-Sforza, L. L. (1983). The transition to agriculture and some of its consequences. In *How Humans Adapt: A Biocultural Odyssey*. ed. D. J. Ortner, pp. 103–26. Washington, DC: Smithsonian Institute Press.

Ceesay, S. M., Prentice, A. M., Day, K. C. *et al.* (1989). The use of heart rate monitoring in the estimation of energy expenditure: a validation study using whole body calorimetry. *British Journal of Nutrition*, **61**, 175–86.

Chavez, A. & Martinez, C. (1980). Effects of maternal nutrition and dietary supplementation. In *Maternal Nutrition during Pregnancy and Lactation*. ed. H. Aebi and R. Whitehead, pp. 274–84. Bern: Huber.

Chen, A. T. & Falek, A. (1974). Chromosome aberrations in full-term low birthweight neonates. *Human Genetics*, **21**, 13–16.

Chen, L. C., Alauddin Chowdhury, A. K. M. & Huffman, S. L. (1979). Seasonal dimensions of energy protein malnutrition in rural Bangladesh: the role of agriculture, dietary practices, and infection. *Ecology of Food and Nutrition*, **8**, 175–87.

Chen Hengli (1958). *The Study of 'Supplementary Agricultural Book' (Bunongshu Yanjiu)*. Beijing: Zhonghua Shujue.

Chibnick, M. & de Jong, W. (1989). Agricultural labor organisation in Ribereno communities of the Peruvian Andes. *Ethology*, **28**, 75–95.

Chisholm, M. (1962). *Rural Settlement and Land Use*. London: Hutchinson.

Chiswick, M. L. (1985). Intrauterine growth retardation. *British Medical Journal*, **291**, 845–8.

Chowdhury, A. K. M. A., Huffman, S. L. & Chen, L. C. (1981). Agriculture and nutrition in Matlab Thana, Bangladesh. In *Seasonal Dimensions to Rural Poverty*. ed. R. Chambers, R. Longhurst & A. Pacey, pp. 52–61. London: Frances Pinter.

Chumlea, W. C., Guo, S. S., Bellisari, A., Baumgartner, R. N. & Siervogel, R. M.

(1994). Reliability for multiple frequency bioelectrical impedance. *American Journal of Human Biology*, **6**, 195–202.

Clark, H. D. & Hoffer, L. J. (1991). Reappraisal of the resting metabolic rate of normal young men. *American Journal of Clinical Nutrition*, **53**, 21–6.

Clark, W. C. (1971). *Place and People: An Ecology of a New Guinean Community*. Berkeley: University of California Press.

Cleland, J. G. & Sathar, Z. A. (1984). The effect of birth spacing on child mortality in Pakistan. *Population Studies*, **38**, 401–10.

Clutton-Brock, T. H. & Harvey, P. H. (1980). Primates, brains and ecology. *Journal of Zoology*, **190**, 309–23.

Cobbett, W. (1830). *Rural Rides*. Reprinted 1967. London: Penguin.

Cody, M. L. (1974). Optimisation in ecology. *Science*, **183**, 1156–64.

Coghill, B. (1982). Ranking anthropometric indicators using mortality in rural Bangladeshi children. M.Sc. thesis, Cornell University.

Coghill, B. (1987). *Seasonal Influences in South-West Kenya*. Washington: International Food Policy Research Institute Research Report.

Cohen, B. (1993). Fertility levels, differentials and trends. In *Demographic Change in Sub-Saharan Africa*. ed. K. A. Foote, K. H. Hill & L. G. Martin, pp. 8–67. Washington, DC: National Academy Press.

Cole, T. J. (1989). The British, American NCMS, and Dutch weight standards compared using the LMS method. *American Journal of Human Biology*, **1**, 397–408.

Cole, T. J. (1991). Sampling, study size, and power. In *Design Concepts in Nutritional Epidemiology*. ed. B. M. Margetts & M. Nelson, pp. 53–78. Oxford: Oxford University Press.

Cole, T. J. (1993). Seasonal effects on physical growth and development. In *Seasonality and Human Ecology*. ed. S. J. Ulijaszek & S. S. Strickland, pp. 89–106. Cambridge: Cambridge University Press.

Coleman, D. & Salt, J. (1992). *The British Population*. Oxford: Oxford University Press.

Collins, K. J. & Spurr, G. B. (1990). Energy expenditure and habitual activity. In *Handbook of Methods for the Measurement of Work Performance, Physical Fitness and Energy Expenditure in Tropical Populations*. ed. K. J. Collins, pp. 81–93. Paris: International Union of Biological Science.

Conklin, H. C. (1957). Hanunoo agriculture, a report on an integral system of shifting cultivation in the Philippines. *Forestry Development Paper No. 12*. Rome: Food and Agriculture Organisation of the United Nations.

Coward, W. A. (1988). The $^2H_2{}^{18}O$ technique: principles and practice. *Proceedings of the Nutrition Society*, **47**, 209–18.

Coward, W. A. & Prentice, A. M. (1985). Isotope method for the measurement of carbon dioxide production rate in man. *American Journal of Clinical Nutrition*, **41**, 659–61.

Coward, W. A., Goldberg, G. & Prentice, A. M. (1992). Energy balance in lactation. In *Mechanisms Regulating Lactation and Infant Nutrient Utilization*. ed. M. F. Picciano & B. Lonnerdal, pp. 65–76. New York: Wiley Liss.

Crittenden, R. & Baines, J. (1986). The seasonal factors influencing child malnutrition on the Nembi plateau, Papua New Guinea. *Human Ecology*, **14**, 191–223.

Cumming, D.C., Wheeler, G.D. & Harber, V.J. (1994). Physical activity, nutrition, and reproduction. In *Human Reproductive Ecology. Interactions of Environment, Fertility, and Behavior*. ed. K.L. Campbell & J.W. Wood, pp. 55–76. New York: New York Academy of Sciences.

Dale, E., Gerlach, D.H. & Wilhite, A.L. (1979). Menstrual dysfunction in distance runners. *Obstetrics and Gynecology*, **54**, 42–52.

Daly, J.M., Heymsfield, S.B., Head, C.A. *et al.* (1985). Human energy requirements: overestimation by widely used prediction equation. *American Journal of Clinical Nutrition*, **42**, 1170–4.

Danford, L.C., Schoeller, D.A. & Kushner, R.F. (1992). Comparison of two bioelectrical analysis models for total body water measurement in children. *Annals of Human Biology*, **19**, 603–7.

Danforth, E. (1985). Hormonal adaptation to over- and underfeeding. In *Substrate and Energy Metabolism*. ed. J.S. Garrow & D. Halliday, pp. 155–66. London: John Libbey.

Danforth, E. & Burger, A.G. (1989). The impact of nutrition on thyroid hormone physiology and action. *Annual Reviews in Nutrition*, **9**, 201–27.

Dauncey, M.J. (1990). Thyroid hormones and thermogenesis. *Proceedings of the Nutrition Society*, **49**, 203–15.

Davies, D.P., Platts, P., Pritchard, J. & Wilkinson, P. (1979). Nutritional status of light-for-date infants at birth and its influence on early postnatal growth. *Archives of Disease in Childhood*, **54**, 703–6.

Dazhong, W. & Pimentel, D. (1986). Seventeenth century organic agriculture in China: II. Energy flows through an agroecosystem in Jiaxing Region. *Human Ecology*, **14**, 15–28.

Delvoye, P., Delogne-Desnoeck, J. & Robyn, C. (1980). Hyperprolactinaemia during prolonged lactation: evidence for anovulatory cycles and inadequate corpus luteum. *Clinical Endocrinology*, **13**, 243–7.

Deurenberg, P., van der Kooy, K., Leenen, R., Westrate, J.A. & Seidell, J.C. (1991). Sex age specific prediction formulas for estimating body composition from bioelectrical impedance: a cross-validation study. *International Journal of Clinical Nutrition*, **15**, 17–25.

Dewey, K.G., Heinig, M.J. & Nommsen, L.A. (1993). Maternal weight-loss patterns during prolonged lactation. *American Journal of Clinical Nutrition*, **58**, 162–6.

Dorman, S. (1925). *Pygmies and Bushmen of the Kalahari*. London: Seeley Service & Company.

Douglas, C.G. (1911). A method for determining the total respiratory exchange in man. *Journal of Physiology* (London), **42**, xvii–xviii.

Dugdale, A.E. (1986). Evolution and infant feeding. *Lancet* **1**, 670–3.

Dugdale, A.E. & Payne, P.R. (1986). Modelling seasonal changes in energy balance. In: *Proceedings of the XIII International Congress of Nutrition*. ed. T.G. Taylor and N.K. Jenkins, pp. 141–4. London: John Libbey.

Dunbar, R.I.M. (1992). Neocortex size as a constraint on group size in primates. *Journal of Human Evolution*, **20**, 469–93.

Durnin, J.V.G.A. (1987). Energy requirements of pregnancy: an integration of the longitudinal data from the five country study. *Lancet*, **2**, 1131–4.

Durnin, J.V.G.A. (1988). The energy requirements of pregnancy and lactation. In

Chronic Energy Deficiency: Consequences and Related Issues. ed. B. Schurch & N. S. Scrimshaw, pp. 135–52. Lausanne: Nestle Foundation.

Durnin, J. V. G. A. (1990a). Methods to assess physical activity and the energy expended for it by infants and children. In *Activity, Energy Expenditure and Energy Requirements of Infants and Children.* ed. B. Schurch & N. S. Scrimshaw, pp. 45–55. Lausanne: International Dietary Energy Consultancy Group.

Durnin, J. V. G. A. (1990b). Low energy expenditures in free-living populations. *European Journal of Clinical Nutrition,* **44**(Supplement 1), 95–102.

Durnin, J. V. G. A. & Ferro-Luzzi, A. (1982). Conducting and reporting studies on human energy intake and output: suggested standards. *Human Nutrition: Clinical Nutrition,* **37A**, 141–4.

Durnin, J. V. G. A. & Passmore, R. (1967). *Energy, Work and Leisure.* London: Heinemann.

Durnin, J. V. G. A. & Womersley, J. (1974). Body fat assessed from total density and its estimation from skinfold thickness: measurements on 481 men and women aged from 16 to 72 years. *British Journal of Nutrition,* **32**, 77–97.

Durnin, J. V. G. A., Drummond, S. & Satyanarayana, K. (1990). A collaborative EEC study on seasonality and marginal nutrition: The Glasgow-Hyderabad (S. India) study. *European Journal of Clinical Nutrition,* **44**(Supplement 1), 19–29.

Dwyer, P. D. (1983). Etolo hunting performance and energetics. *Human Ecology,* **11**, 145–74.

Edholm, O. G., Fletcher, J. G., Widdowson, E. M. & McCance, R. A. (1955). The energy expenditure and food intake of individual men. *British Journal of Nutrition,* **9**, 286–300.

Edholm, O. G., Adam, J. M., Heavy, M. J. R., Wolff, H. S., Goldsmith, R. & Best, T. W. (1970). Food intake and energy expenditure of army recruits. *British Journal of Nutrition,* **24**, 1091–107.

Edmundson, W. C. & Edmundson, S. A. (1989). Energy balance, nutrient intake and discretionary activity in a South Indian village. *Ecology of Food and Nutrition,* **22**, 253–65.

Elia, M. (1992). Organ and tissue contribution to metabolic rate. In *Energy Metabolism: Tissue Determinants and Cellular Corollaries,* ed. J. M. Kinney & H. N. Tucker, pp. 61–79. New York: Raven Press.

Ellen, R. (1978). *Nuaulu Settlement and Ecology: An Approach to the Environmental Relations of an Eastern Indonesian Community.* The Hague: Martinus Nijhoff.

Ellen, R. (1982). *Environment, Subsistence and System.* Cambridge: Cambridge University Press.

Ellison, P. T. (1991). Reproductive ecology and human fertility. In *Applications of Biological Anthropology to Human Affairs.* ed. G. W. Lasker & C. G. N. Mascie-Taylor, pp. 14–54. Cambridge: Cambridge University Press.

Ellison, P. T. (1994). Salivary steroids and natural variation in human function. In *Human Reproductive Ecology. Interactions of Environment, Fertility, and Behaviour.* ed. K. L. Campbell & J. W. Wood, pp. 287–98. New York: New York Academy of Sciences.

Ellison, P. T. & Lager, C. (1985). Exercise-induced menstrual disorders. *New England Journal of Medicine,* **313**, 825–6.

Ellison, P. T., Peacock, N. R. & Lager, C. (1989). Ecology and ovarian function among Lese women of the Ituri forest, Zaire. *American Journal of Physical Anthropology*, **78**, 519–26.

Epstein, T. S. (1962). *Economic Development and Social Change in South India*. Manchester: Manchester University Press.

Evans-Pritchard, E. (1940). *The Nuer*. Oxford: Clarendon Press.

Eveleth, P. B. & Tanner, J. M. (1990). *Worldwide Variation in Human Growth*, 2nd edn. Cambridge: Cambridge University Press.

Fage, A. (1993). The effects of training on menstrual function in Cambridge University oarswomen. Undergraduate dissertation, University of Cambridge, UK.

Falkner, F., Holzgreve, W. & Schloo, R. H. (1994). Prenatal influences on postnatal growth: overview and pointers for needed research. *European Journal of Clinical Nutrition*, **48**(Supplement 1), S15–24.

FAO/WHO/UNU (1985). Energy and Protein Requirements. *Technical Report Series No. 72*. Geneva: World Health Organisation.

Fauno, P., Kalund, S. & Kaustrup, I. L. (1991). Menstrual patterns in elite Danish swimmers. *European Journal of Applied Physiology*, **62**, 36–9.

Fei Hsiao-t'ung & Cahng Chih-i (1945). *Earthbound China: a Study of Rural Economy in Yunnan*. Chicago: University of Chicago Press.

Ferguson, A. G. (1987). Some aspects of birth seasonality in Kenya. *Social Science and Medicine*, **25**, 793–801.

Ferguson, E. L., Gibson, R. S., Ounpuu, S. & Sabry, J. H. (1989). The validity of the 24 hour recall for estimating the energy and selected nutrient intakes of a group of rural Malawian preschool children. *Ecology of Food and Nutrition*, **23**, 273–85.

Ferguson, E. L., Gibson, R. S. & Opare-Obisaw, C. (1994). The relative validity of the repeated 24 h recall for estimating energy and selected nutrient intakes of rural Ghanaian children. *Europen Journal of Clinical Nutrition*, **48**, 241–52.

Ferguson, R. B. (1989). Ecological consequences of Amazonian warfare. *Ethnology*, **28**, 249–64.

Ferin, M., van Vugt, D. & Wardlaw, S. (1984). The hypothalamic control of the menstrual cycle and the role of endogenous opioid peptides. Recent progress in *Hormone Research*, **40**, 441–85.

Ferro-Luzzi, A. (1982). Meaning and constraints of energy-intake studies in free-living populations. In *Energy and Effort*. ed. G. A. Harrison, pp. 115–37. London: Taylor and Francis.

Ferro-Luzzi, A. (1985). Work capacity and productivity in long-term adaptation to low energy intakes. In *Nutritional Adaptation in Man*. ed. K. Blaxter & J. C. Waterlow, pp. 61–9. London: John Libbey.

Ferro-Luzzi, A. (1990a). Social and public health issues in adaptation to low energy intakes. *American Journal of Clinical Nutrition*, **51**, 309–15.

Ferro-Luzzi, A. (1990b). Seasonal energy stress in marginally nourished rural women: interpretation and integrated conclusions of a multicentre study in three developing countries. *European Journal of Clinical Nutrition*, **44**(Supplement 1), 41–6.

Ferro-Luzzi, A. & Branca, F. (1993). Nutritional seasonality: the dimensions of the problem. In *Seasonality and Human Ecology*. ed. S. J. Ulijaszek & S. S.

Strickland, pp. 149–65. Cambridge: Cambridge University Press.

Ferro-Luzzi, A., Norgan, N. G. & Paci, C. (1981). An evaluation of protein and energy intakes in some New Guinean households. *Nutrition Reports International*, **24**, 153–63.

Ferro-Luzzi, A., Scaccini, C., Taffese, S., Aberra, B. & Demeke, T. (1990). Seasonal energy deficiency in Ethiopian rural women. *European Journal of Clinical Nutrition*, **44**(Supplement 1), 7–18.

Ferro-Luzzi, A., Sette, S., Franklin, M. & James, W. P. T. (1992). A simplified approach of assessing adult chronic energy deficiency. *European Journal of Clinical Nutrition*, **46**, 173–86.

Fibiger, W. & Singer, G. (1989). Biochemical assessment and differentiation of mental and physical effort. *Work and Stress*, **3**, 237–47.

Fink, A. E., Fink, G., Wilson, H., Bennie, J., Carroll, S. & Dick, H. (1992). Lactation, nutrition and fertility and the secretion of prolactin and gonadotrophins in Mopan Mayan women. *Journal of Biosocial Science*, **24**, 35–52.

Fjeld, C. R., Freundt-Thurne, J. & Schoeller, D. A. (1990). Total body water measured by 180 dilution and bioelectrical impedance in well and malnourished children. *Pediatric Research*, **27**, 98–102.

Flatt, J. P. (1985). Energetics of intermediary metabolism. In *Substrate and Energy Metabolism in Man*. ed. J. S. Garrow & D. Halliday, pp. 58–69. London: John Libbey.

Fleagle, J. G. (1992). Primate location and posture. In *Encyclopedia of Human Evolution*. ed. S. Jones, R. Martin & D. Pilbeam, pp. 75–9. Cambridge: Cambridge University Press.

Foley, R. A. (1987). *Another Unique Species. Patterns in Human Evolutionary Ecology*. Harlow, Essex: Longman.

Foley, R. A. (1992). Evolutionary ecology of fossil hominids. In *Evolutionary Ecology and Human Behavior*. ed. E. A. Smith & B. Winterhalder, pp. 131–64. New York: Aldine de Gruyter.

Foley, R. A. & Lee, P. C. (1989). Finite social space, evolutionary pathways and reconstructing hominid behavior. *Science*, **243**, 901–6.

Foley, R. A. & Lee, P. C. (1991). Ecology and energetics of encephalization in hominid evolution. *Philosophical Transactions of the Royal Society of London, Series B*, **334**, 223–32.

Forbes, G. (1987). *Human Body Composition. Growth, Aging, Nutrition, and Activity*. Berlin: Springer-Verlag.

Forbes, G. B. (1989). Changes in body composition. In *Report of the 98th Ross Conference on Pediatric Research*, pp. 112–18. Columbus, OH: Ross Laboratories.

Forsum, E., Sadurskis, A. & Wager, J. (1988). Resting metabolic rate and body composition of healthy Swedish women during pregnancy. *American Journal of Clinical Nutrition*, **47**, 942–7.

Fox, H. S., Campbell, V. S. & Lovell, H. G. (1968). A comparison of dietary data obtained in Jamaica by twenty-four-hour recall and by weighing. *Archives of Latinamerican Nutrition*, **18**, 81–97.

Fox, R. H. (1953). A study of the energy expenditure of Africans engaged in various activities, with special reference to some environmental and physiological factors which may influence the efficiency of their work. PhD thesis. London:

University of London.

Fraser, D. R. (1988). Nutritional growth retardation: experimental studies with special reference to calcium. In *Linear Growth Retardation in Less Developed Countries.* ed. J. C. Waterlow, pp. 127–35. New York: Raven Press.

Freeman, D. (1970). Report on the Iban. *London School of Economics Monographs on Social Anthropology No. 41.* London: Athlone Publications.

Frigerio, C., Schutz, Y., Whitehead, R. & Jecquier, E. (1992). Postprandial thermogenesis in lactating and non-lactating women from The Gambia. *European Journal of Clinical Nutrition,* **46**, 7–13.

Frisancho, A. R. (1976). Growth and morphology at high altitude. In *Man in the Andes. A Multidisciplinary Study of High-altitude Quechua.* ed. P. T. Baker & M. A. Little, pp. 180–207. Stroudsburg, Pennsylvania: Dowden, Hutchinson & Ross.

Frisancho, A. R. (1990). *Anthropometric Standards for the Assessment of Growth and Nutritional Status.* Ann Arbor: University of Michigan Press.

Frisancho, A. R. (1993). *Human Adaptation and Accommodation.* Ann Arbor: University of Michigan Press.

Galle, P. C., Freeman, E. W., Galle, M. G., Huggins, G. R. & Sondheimer, S. J. (1983). Physiologic and psychologic profiles in a survey of women runners. *Fertility and Sterility,* **39**, 633–9.

Gallo, P. G. & Mestriner, M. F. (1980). Growth of children in Somalia. *Human Biology,* **52**, 547–62.

Garby, L., Kurzer, M. S., Lammert, O. & Nielsen, E. (1987). Energy expenditure during sleep in men and women: evaporative and sensible heat losses. *Human Nutrition: Clinical Nutrition,* **41C**, 225–33.

de Garine, I. (1993). Culture, seasons and stress in two traditional African cultures (Massa and Mussey). In *Seasonality and Human Ecology.* ed. S. J. Ulijaszek & S. S. Strickland, pp. 184–201. Cambridge: Cambridge University Press.

de Garine, I. & Harrison, G. A. (1988). Discussion and conclusions. In *Coping with Uncertainty in Food Supply.* ed. I. de Garine & G. A. Harrison, pp. 469–75. Oxford: Oxford University Press.

de Garine, I. & Koppert, G. (1988). Coping with seasonal fluctuations in food supply among savanna populations: the Massa and Mussey of Chad and Cameroon. In *Coping with Uncertainty in Food Supply.* ed. I. de Garine & G. A. Harrison, pp. 210–259. Oxford: Oxford University Press.

de Garine, I. & Koppert, S. (1990). Social adaptation to season and uncertainty in food supply. In *Diet and Disease in Traditional and Developing Societies.* ed. G. A. Harrison & J. C. Waterlow, pp. 240–89. Cambridge: Cambridge University Press.

Garland, T., Geiser, F. & Baudinette, R. V. (1988). Comparative locomotor performance of marsupial and placental mammals. *Journal of Zoology,* **215**, 505–22.

Garrow, J. S. (1981). *Treat Obesity Seriously.* Edinburgh: Churchill Livingstone.

Garrow, J. S. (1985). Responses to overnutrition. In *Nutritional Adaptation in Man.* ed. K. Blaxter & J. C. Waterlow, pp. 105–9. London: John Libbey.

Gessain, M. (1978). Poids individuels saisonniers chez les Basssari du Sénégal Oriental. *Bulletin et Mémoires de la Société d'Anthropologie de Paris,* **5**, 149–55.

Gibson, R. S. (1990). *Principles of Nutritional Assessment*. Oxford: Oxford University Press.

Glass, A. R., Yahiro, J. A., Deuster, P. A., Vigersky, E. A., Kyle, S. B. & Schoomaker, E. B. (1987). Amenorrhoea in Olympic marathon runners. *Fertility and Sterility*, **48**, 740–5.

Goldberg, G. R., Prentice, A. M., Davies, H. L. & Murgatroyd, P. R. (1988). Overnight and basal metabolic rates in men and women. *European Journal of Clinical Nutrition*, **42**, 137–44.

Goldberg, G. R., Davies, H. L., Prentice, A. M., Coward, W. A. & Sawyer, M. (1991). How is the energy budget balanced in well-nourished lactating women? *Proceedings of the Nutrition Society*, **50**, 8A.

Goldberg, G. R., Prentice, A. M., Coward, W. A. *et al.* (1993). Longitudinal assessment of energy expenditure in pregnancy by the doubly labeled water method. *American Journal of Clinical Nutrition*, **57**, 494–505.

Golden, M. H. N. (1988). The role of individual nutrient deficiencies in growth retardation of children as exemplified by zinc and protein. In *Linear Growth Retardation in Less Developed Countries*. ed. J. C. Waterlow, pp. 143–60. New York: Raven Press.

Golden, M. H. N. (1994). Is complete catch-up possible for stunted malnourished children? *European Journal of Clinical Nutrition*, **48**(Supplement 1), S58–70.

Golden, M. H. N., Waterlow, J. C. & Picou, D. (1977). The relationship between dietary intake, weight change, nitrogen balance and protein turnover in man. *American Journal of Clinical Nutrition*, **30**, 1345–8.

Gordon, K. (1987). Evolutionary perspectives on human diet. In *Nutritional Anthropology*. ed. F. E. Johnston, pp. 3–40. New York: Alan R. Liss.

Grantham-McGregor, S., Gardner, J. M. M., Walker, S. & Powell, C. (1990). The relationship between undernutrition, activity levels and development in young children. In *Activity, Energy Expenditure and Energy Requirements of Infants and Children*. ed. B. Schurch & N. S. Scrimshaw, pp. 361–83. Lausanne: International Dietary Energy Consultancy Group.

Gray, R. H., Campbell, O. M., Apelo, R. *et al.* (1990). Risk of ovulation during lactation. *Lancet*, 335, 25–9.

Greenfield, H. & Southgate, D. A. T. (1992). *Food composition data. Production management and use*. London: Elsevier.

Grove, E. W., Olmstead, W. H. & Koenig, K. (1929). The effect of diet and catharsis on the lower volatile fatty acids in the stools of normal men. *Journal of Biological Chemistry*, **85**, 127–36.

Guillermo-Tuazon, M. A., Barba, C. V. C., van Raaij, J. M. A. & Hautvast, J. G. A. J. (1992). Energy intake, energy expenditure, and body composition of poor rural Philippine women throughout the first 6 months of lactation. *American Journal of Clinical Nutrition*, **56**, 874–80.

Guo, S., Roche, A. F., Chumlea, W. M. C., Miles, D. S. & Pohlman, R. L. (1987). Body composition predictions from bioelectric impedance. *Human Biology*, **59**, 221–33.

Guo, S., Roche, A. F. & Houtkooper, L. (1989). Fat-free mass in children and young adults predicted from bioelectric impedance and anthropometric variables. *American Journal of Clinical Nutrition*, **50**, 435–43.

Guptill, K., Berendes, H., Forman, M. R. *et al.* (1990). Seasonality of births among bedouin arabs residing in the Negev desert of Israel. *Journal of Biosocial Science*, **22**, 493–506.

de Guzman, P. E. (1981). Energy Allowances for the Philippine Population. *Proceedings of a Workshop on Energy Expenditure under Field Conditions*, Prague.

de Guzman, P. E. (1984). Energy expenditure, dietary intake, and patterns of daily activity among various occupation groups in the Philippines. In *Protein-energy-requirement Studies in Developing Countries: Results of International Research*. ed. W. Rand, R. Uauy & N. S. Scrimshaw, pp. 187–92. Tokyo: United Nations University.

de Guzman, P. E., Dominguez, S. R., Kalaw, J. M., Basconcillo, R. O. & Santos, V. F. (1974). A study of the energy expenditure, dietary intake, and pattern of daily activity among various occupational groups. I. Laguna rice farmers. *Philippine Journal of Science*, **103**, 53–65.

Haas, J. D. & Pelletier, D. L. (1989). Nutrition and human population biology. In *Human Population Biology*, ed. M. A. Little & J. D. Haas, pp. 152–67. Oxford: Oxford University Press.

Haas, J. D., Balcazar, H. & Caulfield, L. (1987). Variation in early neonatal mortality for different types of fetal growth retardation. *American Journal of Physical Anthropology*, **73**, 467–73.

Haeckel, E. (1868). *Naturliche Schopfungsgeschichte*. Berlin: Georg Reimer.

Haggarty, P. & McGaw, B. A. (1988). Non-restrictive methods for measuring energy expenditure. *Proceedings of the Nutrition Society*, **47**, 365–74.

Harris, D. R. (1981). The prehistory of human subsistence: a speculative outline. In *Food, Nutrition and Evolution*, ed. D. N. Waltcher & N. Kretchmer, pp. 15–35. New York: Masson Publications.

Harris, J. A. & Benedict, F. G. (1919). A biometric study of basal metabolism in man. *Publication No. 279*. Washington: Carnegie Institution.

Harris, M. (1971). *Culture, Man and Nature: an Introduction to General Anthropology*. New York: Crowell.

Harrison, G. A. (1982). Preface. In *Energy and Effort*. ed. G. A. Harrison, pp. vii–ix. London: Taylor and Francis.

Harrison, G. A. (1993). *Human Adaptation*. Oxford: Oxford University Press.

Harrison, G. A., Weiner, J. S., Tanner, J. M. & Barnicot, N. A. (1964). *Human Biology. An Introduction to Human Evolution, Variation, Growth, and Ecology*. Oxford: Oxford University Press.

Harrison, G. A., Brush, G., Almedom, A. & Jewell, T. (1990). Short-term variations in stature growth in Ethiopian and English children. *Annals of Human Biology*, **17**, 407–16.

Hartman, M. L., Iranmanesh, A., Thorner, M. O. & Veldhuis, J. D. (1993). Evaluation of pulsatile patterns of growth hormone release in humans: a brief review. *American Journal of Human Biology*, **5**, 603–14.

Harvey, P. H., Martin, R. D. & Clutton-Brock, T. H. (1987). Life histories in comparative perspective. In *Primate Societies*. ed. B. B. Smuts, D. L. Cheney, R. M. Seyfarth, R. W. Wrangham & T. T. Struhsaker, pp. 181–96. Chicago: University of Chicago Press.

Haswell, M. R. (1953). Economics of agriculture in a savannah village. *Colonial Research Study No. 8*. London: Her Majesty's Stationery Office.

Haswell, M. R. (1981). Food consumption in relation to labour output. In *Seasonal Dimensions of Rural Poverty*. ed. R. Chambers, R. Longhurst & A. Pacey, pp. 38–41. London: Frances Pinter.

Hauck, H. M., Thorangkul, D. & Rajatasilpin, A. (1960). Growth in height and weight of elementary school children. *Journal of Tropical Pediatrics*, 6, 84–91.

Hawkes, K., Hill, K. & O'Connell, J. F. (1982). Why hunters gather: optimal foraging and the Ache of Eastern Paraguay. *American Ethnologist*, 9, 379–98.

Hawley, A. H. (1950). *Human Ecology. A Theory of Community Structure*. New York: The Ronald Press Company.

Hayden, B. (1981). Subsistence and ecological adaptations of modern hunter–gatherers. In *Omnivorous Primates*. ed. R. S. O. Harding & G. Teleki, pp. 344–421. New York: Colombia University Press.

Hayter, J. E. & Henry, C. J. K. (1993). Basal metabolic rate in human subjects migrating between tropical and temperate regions: a longitudinal study and review of previous work. *European Journal of Clinical Nutrition*, 47, 724–34.

Headland, T. N. (1987). The wild yam question: how well could independent hunter–gatherers live in a tropical rain forest ecosystem? *Human Ecology*, 15, 463–91.

Heini, A., Schutz, Y. & Jequier, E. (1992). Twenty-four-hour energy expenditure in pregnant and nonpregnant Gambian women, measured in a whole-body indirect calorimeter. *American Journal of Clinical Nutrition*, 55, 1078–85.

Heitmann, B. L. (1990). Prediction of body water and fat in adult Danes from measurement of electrical impedance. A validation study. *International Journal of Obesity*, 14, 789–802.

Heitmann, B. L. (1994). Impedance: a valid method in assessment of body composition? *European Journal of Clinical Nutrition*, 48, 228–40.

Henry, C. J. K. & Rees, D. D. (1988). A preliminary analysis of basal metabolic rate and race. In *Comparative Nutrition*. ed. I. Macdonald, pp. 149–59. London: John Libbey.

Henry, C. J. K. & Rees, D. G. (1991). New predictive equations for the estimation of basal metabolic rate in tropical peoples. *European Journal of Clinical Nutrition*, 45, 177–85.

Herbert, J., Moore, G., de la Riva, C. & Watts, F. (1986). Endocrine responses and examination anxiety. *Biological Psychology*, 22, 215–26.

Heywood, P. (1982). The functional significance of malnutrition: growth and prospective risk of death in the Highlands of Papua New Guinea. *Journal of Food and Nutrition*, 39, 13–19.

Hilderbrand, K. (1985). Assessing the components of seasonal stress amongst Fulani of the Seno-Mango, Central Mali. In *Population, Health and Nutrition in the Sahel. Issues in the Welfare of Selected West African Communities*. ed. A. G. Hill, pp. 208–87. London: Routledge & Kegan Paul.

Hill, K. (1982). Hunting and human evolution. *Journal of Human Evolution*, 11, 521–44.

Hill, K. & Kaplan, H. (1988a). Tradeoffs in male and female reproductive strategies among the Ache: part 1. In *Human Reproductive Behaviour*. ed. L. Betzig, M. Borgerhoff Mulder & P. Turke, pp. 277–90. Cambridge: Cambridge University Press.

Hill, K. & Kaplan, H. (1988b). Tradeoffs in male and female reproductive strategies among the Ache: part 2. In *Human Reproductive Behaviour*. ed. L. Betzig, M. Borgerhoff Mulder & P. Turke, pp. 291–306. Cambridge: Cambridge University Press.

Hill, K., Kaplan, H., Hawkes, K. & Hurtado, A. (1984). Seasonal variance in the diet of Ache hunter–gatherers in eastern Paraquay. *Human Ecology*, **12**, 145–80.

Hill, K., Kaplan, H., Hawkes, K. & Hurtado, A. (1985). Men's time allocation to subsistence work among the Ache of Eastern Paraquay. *Human Ecology*, **13**, 29–47.

Hitchcock, R. K. (1988). Settlement, seasonality, and subsistence stress among the Tyua of Northern Botswana. In *Coping with Seasonal Constraints*. ed. R. Huss-Ashmore, J. J. Curry & R. K. Hitchcock, pp. 64–85. Philadelphia: MASCA Research Papers in Science and Archaeology.

Hodgdon, J. A. & Fitzgerald, P. I. (1987). Validity of impedance at various levels of fatness. *Human Biology*, **59**, 281–98.

Hoff, C. (1968). Reproduction and viability in a highland Peruvian Indian population. In *High Altitude Adaptation in a Peruvian Community*. *Occasional Papers in Anthropology*, No. 1. Pennsylvania: Department of Anthropology, Pennsylvania State University.

Holloszy, J. O. & Coyle, E. F. (1984). Adaptations of skeletal muscle to endurance exercise and their metabolic consequences. *Journal of Applied Physiology*, **56**, 831–8.

Horton, E. S. (1983). Introduction: an overview of the assessment and regulation of energy balance in humans. *American Journal of Clinical Nutrition*, **38**, 972–77.

Houtkooper, L. B., Lohman, T. G., Going, S. B. & Hall, M. C. (1989). Validity of bioelectric impedance for body composition assessment in children. *Journal of Applied Physiology*, **66**, 814–21.

Houtkooper, L. B., Going, S. B., Lohman, T. G., Roche, A. F. & van Loan, M. (1992). Bioelectrical impedance estimation of fat-free body mass in children and youth: a cross-validation study. *Journal of Applied Physiology*, **72**, 366–73.

Howie, P. W. & McNeilly, A. S. (1982). Effect of breastfeeding patterns on human birth intervals. *Journal of Reproductive Fertility*, **65**, 545–57.

Hoyle, B., Yunus, M. D. & Chen, L. C. (1980). Breast-feeding and food intake among children with acute diarrheal disease. *American Journal of Clinical Nutrition*, **33**, 2365–71.

Hubert, W. & de Jong-Meyer, R. (1989). Emotional stress and saliva cortisol response. *Journal of Clinical Chemistry and Clinical Biochemistry*, **27**, 235–7.

Hull, V. J. (1983). The Ngaglik study: an inquiry into birth interval dynamics and maternal and child health in rural Java. *World Health Statistics Quarterly*, **36**, 100–18.

Humphrey, N. K. (1976). The social function of intellect. In *Growing Points in Ethology*, ed. P. P. G. Bateson & R. A. Hinde, pp. 303–17. Cambridge: Cambridge University Press.

Humphrey, S. J. E. & Wolff, H. S. (1977). The Oxylog. *Journal of Physiology*, **267**, 12P.

Huss-Ashmore, R. A. (1988). Seasonal patterns of birth and conception in rural highland Lesotho. *Human Biology*, **60**, 493–506.

Huss-Ashmore, R. A. (1993). Agriculture, modernisation and seasonality. In

Seasonality and Human Ecology, ed. S.J. Ulijaszek & S.S. Strickland, pp. 202–19. Cambridge: Cambridge University Press.

Huss-Ashmore, R.A. (1995). Evaluation of energy intake methods. *American Journal of Human Biology*, in press.

Huss-Ashmore, R.A. & Goodman, J.L. (1988). Seasonality of work, weight, and body composition for women in highland Lesotho. In *Coping with Seasonal Constraints*. ed. R. Huss-Ashmore, J.J. Curry & R.K. Hitchcock, pp. 29–44. Philadelphia: MASCA Research Papers in Science and Archaeology.

Huss-Ashmore, R.A. & Thomas, R.B. (1988). A framework for analysing uncertainty in highland areas. In *Coping with Uncertainty in Food Supply*. ed. I. de Garine and G.A. Harrison, pp. 452–68. Oxford: Oxford University Press.

Huss-Ashmore, R.A., Goodman, J.L., Sibiya, T.E. & Stein, T.P. (1989). Energy expenditure of young Swazi women as measured by the doubly-labelled water method. *European Journal of Clinical Nutrition*, **43**, 737–48.

Hyndman, D.C. (1979). Wopkaimin subsistence: cultural ecology in the New Guinea highland fringe. PhD thesis. Brisbane: University of Queensland.

Hyndman, D.C., (1989). Gender in the diet and health of the Wopkaimin. In *A Continuing Trial of Treatment. Medical Pluralism in Papua New Guinea*. ed. S. Frankel & G. Lewis, pp. 249–75. Dordrecht: Kluwer Academic.

Hyndman, D.C., Ulijaszek, S.J. & Lourie, J.A. (1989). Variability in subsistence ecology and body physique among peoples of the Fly River in Papua New Guinea. *American Journal of Physical Anthropology*, **79**, 89–101.

Hytten, F.E. (1980). Nutrition. In *Clinical Physiology in Obstetrics*. ed. F.E. Hytten & G. Chamberlain, pp. 163–92. Oxford: Blackwell Scientific.

Hytten, F.E. & Leitch, I. (1971). *The Physiology of Human Pregnancy*, 2nd edn. Oxford: Blackwell Scientific.

Illingworth, P.J., Jung, R.T., Howie, P.W., Leslie, P. & Isles, T.E. (1986). Diminution in energy expenditure during lactation. *British Medical Journal*, **292**, 437–41.

Immink, M.D.C. & Viteri, F.E. (1981). Energy intake and productivity of Guatemalan sugar-cane cutters. An empirical test of the efficiency wage hypothesis. Part I and Part II. *Journal of Development Economics*, **9**, 251–87.

Immink, M.D.C., Viteri, F.E., Flores, R. & Torun, B. (1984). Microeconomic consequences of energy deficiency in rural populations in developing countries. In *Energy Intake and Activity*. ed. E. Pollitt and R. Amante, pp. 355–76. New York: A.R. Liss.

Infoods (1986). *International Directory of Food Composition Tables*. Cambridge Massachussetts: International Network of Food Data Systems.

Inoaka, T. (1990). Energy expenditure. In *Population of Human Survival*. ed. R. Ohtsuka & T. Suzuki, pp. 61–7. Tokyo: University of Tokyo Press.

Isaac, G. & Crader, D. (1981). To what extent were early humans carnivores. In *An Archaeological Perspective in Omnivorous Primates*. ed. R. Harding & G. Teleki, pp. 37–109. New York: Columbia University Press.

Isaksson, B. (1980). Urinary nitrogen output as a validity test in dietary surveys. *American Journal of Clinical Nutrition*, **33**, 4–6.

Isley, W.L., Underwood, L.E. & Clemmons, D.R. (1983). Dietary components that regulate serum somatomedin-C concentrations in humans. *Journal of Clinical Investigation*, **71**, 175–82.

Jackson, A. A. (1985). Nutritional adaptation in disease and recovery. In *Nutritional Adaptation in Man*. ed. K. Blaxter & J. C. Waterlow, pp. 111–25. London: John Libbey.

Jackson, A. S. & Pollock, M. L. (1974). Generalised equations for predicting body density of men. *British Journal of Nutrition*, **40**, 497–504.

Jackson, A. S., Pollock, M. L. & Ward, A. (1980). Generalised equations for predicting body density of women. *Medicine and Science in Sports and Exercise*, **12**, 175–82.

Jain, M. G., Harrison, L., Howe, G. R. & Miller, A. B. (1982). Evaluation of a self-administered dietary questionnaire for use in a cohort study. *American Journal of Clinical Nutrition*, **36**, 931–5.

James, W. P. T. (1988). Research relating to energy adaptation in man. In *Chronic Energy Deficiency: Consequences and Related Issues*. ed. B. Schurch & N. S. Scrimshaw, pp. 7–36. Lausanne: Nestle Foundation.

James, W. P. T. & Schofield, E. C. (1990). *Human Energy Requirements. A Manual for Planners and Nutritionists*. Oxford: Oxford University Press.

James, W. P. T., Ferro-Luzzi, A. & Waterlow, J. C. (1988a). Definition of chronic energy deficiency in adults. *European Journal of Clinical Nutrition*, **42**, 969–81.

James, W. P. T., Haggerty, P. & McGaw, B. A. (1988b). Recent progress in studies of energy expenditure: are the new methods providing answers to the old questions? *Proceedings of the Nutrition Society*, **47**, 195–208.

Jelliffe, D. B. & Jelliffe, E. P. (1989). *Assessment of Nutritional Status of the Community*. Oxford: Oxford University Press.

Johnston, F. E., Wainer, H., Thissen, D. & Macvean, R. B. (1976). Hereditary and environmental determinants of growth in height in a longitudinal sample of children and youth of Guatamalan and European ancestry. *American Journal of Physical Anthropology*, **44**, 469–76.

Jolly, C. L. & Gribble, J. N. (1993). The proximate determinants of fertility. In *Demographic Change in Sub-Saharan Africa*. ed. K. A. Foote, K. H. Hill & L. G. Martin, pp. 68–116. Washington, DC: National Academy Press.

Jones, C. D. R. (1989). Energy cost of carrying loads. *European Journal of Clinical Nutrition*, **43**, 881–3.

Jones, C. D. R., Jarjon, M. S., Whitehead, R. G. & Jecquier, E. (1987). Fatness and the energy cost of carrying loads in African women. *Lancet* **ii**, 1331–2.

Kaplan, H. & Hill, K. (1992). The evolutionary ecology of food acquisition. In *Evolutionary Ecology and Human Behavior*. ed. E. A. Smith & B. Winterhalder, pp. 167–201. New York: Aldine de Gruyter.

Karlberg, J. & Albertsson-Wikland, K. (1988). Infancy growth pattern related to growth hormone deficiency. *Acta Paediatrica Scandinavica*, **77**, 385–91.

Kemp, W. B. (1971). The flow of energy in a hunting society. *Scientific American*, **225**, 105–15.

Kennedy, K. I. & Visness, C. M. (1992). Contraceptive efficacy of lactational amenorrhoea. *Lancet*, **339**, 227–30.

Keys, A., Brozek, J., Henschel, A., Michelson, O. & Taylor, H. L. (1950). *The Biology of Human Starvation*. Minneapolis: University of Minnesota Press.

Keys, A., Taylor, H. L. & Grande, F. (1973). Basal metabolism and age of adult man. *Metabolism*, **22**, 579–87.

Khaled, M. A., McCutcheon, M. J., Reddy, S., Pearmen, P. L. Hunter, G. R. &

Weinsier, R. L. (1988). Electrical impedance in assessing human body composition: the BIA method. *American Journal of Clinical Nutrition*, **47**, 789–92.

Kielmann, A. A. & McCord, C. (1978). Weight-for-age as an index of risk of death in children. *Lancet* **i**, 1247–50.

Kinabo, J. L. & Durnin, J. V. G. A. (1990). Thermic effect of food in man: effect of meal composition and energy content. *British Journal of Nutrition*, **64**, 37–44.

King, J. M. & Taitz, L. S. (1985). Catch up growth following abuse. *Archives of Diseases in Childhood*, **60**, 1152–4.

Kirschbaum, C. & Hellhammer, D. (1989). Salivary cortisol in psychobiological research: an overview. Neuropsychobiology, **22**, 150–69.

Kleiber, M. (1961). *The Fire of Life: An Introduction to Animal Energetics.* Huntingdon, New York: Kreiger.

Kofranyi, E. & Michaelis, H. F. (1940). Ein tragbarer apparat zur bestimmung des gastoffwechsels. *Arbietsphysiologie*, **11**, 148–50.

Koishi, H. (1990). Nutritional adaptation of Papua New Guinea Highlanders. *European Journal of Clinical Nutrition*, **44**, 853–85.

Konig, J. (1878). *Chemie der menschlichen Nahrungs- und Genussmittel.* Berlin: Springer Verlag.

Koppert, G. J. A., Dounias, E., Froment, A. & Pasquet, P. (1993). Food consumption in three forest populations of the southern coastal area of Cameroon: Yassa – Mvae – Bakola. In *Tropical Forests, People and Food. Biocultural Interactions and Applications to Development.* ed. C. M. Hladik, A. Hladik, O. F. Linares, H. Pagezy, A. Semple & M. Hadley, pp. 295–310. Paris: UNESCO Publications.

Kramer, M. S. (1987a). Determinants of low birth weight: methodological assessment and meta-analysis. *Bulletin of the World Health Organisation*, **65**, 663–737.

Kramer, M. S. (1987b). Intrauterine growth and gestational duration determinants. *Pediatrics*, **80**, 502–11.

Krasovec, K. (1989). An investigation into the use of maternal arm circumference for nutritional monitoring of pregnant women. ScD. dissertation, Johns Hopkins University.

Kremer, J. A., Borm, G., Schellekens, L. A., Thomas, G. M. G. & Rolland, R. (1991). Pulsatile secretion of luteinising hormone and prolactin in lactating and nonlactating women and the response to naltrexone. *Journal of Clinical Endocrinology and Metabolism*, **72**, 294–300.

Kulkarni, R. N. & Shetty, P. S. (1992). Net mechanical efficiency during stepping in chronically deficient human subjects. *Annals of Human Biology*, **19**, 421–5.

Kushner, R. & Schoeller, D. A. (1986). Estimation of total body water by bioelectrical impedance analysis. *American Journal of Clinical Nutrition*, **44**, 417–24.

Kushner, R. F., Kunigk, A., Alspaugh, M., Andronis, P. T., Leitch, C. A. & Schoeller, D. A. (1990). Validation of bioelectrical-impedance analysis as a measurement of change in body composition in obesity. *American Journal of Clinical Nutrition*, **52**, 219–23.

Lager, C. & Ellison, P. T. (1990). Effect of moderate weight loss on ovarian function assessed by salivary progesterone measurements. *American Journal of Human*

Biology, **2**, 303–12.

Lampl, M. (1993). Evidence of saltatory growth in infancy. *American Journal of Human Biology*, **5**, 641–52.

Lampl, M., Veldhuis, J. D. & Johnson, M. L. (1992). Saltation and stasis: A model of human growth. *Science*, **258**, 801–3.

Langton, J. (1973). Potentialities and problems of adopting a systems approach to the study of change in human geography. *Progress in Geography: International Review of Current Research*, **4**, 125–79.

Lawrence, G. (1992). Pigbel. In *Human Biology in Papua New Guinea*. ed. R. D. Attenborough & M. P. Alpers, pp. 335–44. Oxford: Oxford University Press.

Lawrence, M. & Whitehead, R. G. (1988). Physical activity and total energy expenditure of child-bearing Gambian women. *European Journal of Clinical Nutrition*, **42**, 145–60.

Lawrence, M., Lawrence, F. & Whitehead, R. G. (1984). Maintenance energy cost of pregnancy in rural Gambian women, and influence of dietary status. *Lancet* **ii**, 363–5.

Lawrence, M., Singh, J., Lawrence, F. & Whitehead, R. G. (1985). The energy cost of common daily activities in African women: increased expenditure in pregnancy? *American Journal of Clinical Nutrition*, **42**, 753–63.

Lawrence, M., Coward, W. A., Lawrence, F., Cole, T. J. & Whitehead, R. G. (1987). Fat gain during pregnancy in rural African women: the effect of season and dietary status. *American Journal of Clinical Nutrition*, **45**, 1442–50.

Lawrence, M., Thongprasert, K. & Durnin, J. V. G. A. (1988a). Between-group differences in basal metabolic rates: an analysis of data collected in Scotland, The Gambia and Thailand. *European Journal of Clinical Nutrition*, **42**, 877–91.

Lawrence, M., Singh, F., Lawrence, F. & Whitehead, R. G. (1988b). Energy requirement of pregnancy and lactation: the energy cost of common daily activities in The Gambia. *Annual Report. MRC Dunn Nutrition Unit, UK, and Keneba, Gambia*. Lausanne: Nestlé Foundation.

Lawrence, M., Lawrence, F., Cole, T. J., Coward, W. A., Singh, J. & Whitehead, R. G. (1989). Seasonal pattern of activity and its nutritional consequences in Gambia. In *Seasonal Variability in Third World Agriculture. The Consequences for Food Security*. ed. D. E. Sahn, pp. 47–56. Baltimore: Johns Hopkins Press.

Leach, G. (1976). *Energy and Food Production*. Guildford: IPC Science and Technology Press.

Lechtig, A., Habicht, J. P., Delgado, H., Klein, R. E., Yarbrough, C. & Martorell, R. (1975). Effect of food supplementation during pregnancy on birthweight. *Pediatrics*, **56**, 508–20.

Lee, P. C. (1989). Comparative ethological approaches in modelling hominid behaviour, *Ossa*, **14**, 113–26.

Lee, R. B. (1965). Subsistence Ecology of !Kung Bushmen. PhD thesis. Berkeley: University of California.

Lee, R. B. (1968). What hunters do for a living, or, how to make out on scarce resources. In *Man the Hunter*. ed. R. B. Lee & I de Vore, pp. 30–48. Chicago: Aldine de Gruyter.

Lee, R. B. (1969). !Kung Bushman subsistence: an input–output analysis. In *Contributions to Anthropology: Ecological Essays*. ed. D. Damas, pp. 3–94. Ottawa: Queen's Printers for Canada.

Lee, R. B. & de Vore, I. (eds.) (1968). *Man the Hunter*. Chicago: Aldine de Gruyter.

Leidy, L. E. (1993). The practice of terminal abstinence in Nigeria and Cameroon. *American Journal of Human Biology*, **5**, 565–73.

Leonard, W. R. (1994). Energetics and human population ecology. *American Journal of Human Biology*, **6**, 125–6.

Leonard, W. R. (1995). Energetics and human population ecology. *American Journal of Human Biology*, in press.

Leonard, W. R. & Robertson, M. L. (1992). Nutritional requirements and human evolution: a bioenergetics model. *American Journal of Human Biology*, **4**, 179–95.

Leonard, W. R. & Robertson, M. L. (1994). Evolutionary perspectives on human nutrition: the influence of brain and body size on diet and metabolism. *American Journal of Human Biology*, **6**, 77–88.

Leonard, W. R. & Thomas, R. B. (1989). Biosocial responses to seasonal food stress in highland Peru. *Human Biology*, **61**, 65–85.

van Lerberghe, W. (1990). *Kasongo. Child Mortality and Growth in a Small African Town*. London: Smith Gordon.

Leslie, P. W. & Fry, P. H. (1989). Extreme seasonality of births among nomadic Turkana pastoralists. *American Journal of Physical Anthropology*, **79**, 103–15.

Leung, W-T. W., Butrum, R. R. & Chang, F. H. (1972). *Food Composition Table for Use in East Asia*. Rome: Food and Agriculture Organisation.

Levenstein, S., Prantera, C., Varvo, V. *et al.* (1993). Development of the perceived stress questionnaire: a new tool for psychosomatic research. *Journal of Psychosomatic Research*, **37**, 19–32.

Lewis, O. (1963). *Life in a Mexican Village: Tepoztlan Restudied*. Urbana: University of Illinois Press.

Lifson, N. & McClintock, R. (1966). Theory of use of turnover rates of body water for measuring energy and material balance. *Journal of Theoretical Biology*, **12**, 46–74.

Lincoln, R. J., Boxshall, G. A. & Clark, P. F. (1982). *A Dictionary of Ecology, Evolution and Systematics*. Cambridge: Cambridge University Press.

Lindeman, R. L. (1942). The trophic–dynamic aspect of ecology. *Ecology*, **23**, 399–418.

Little, M. A. (1989). Human biology of African pastoralists. *Yearbook of Physical Anthropology*, **32**, 215–47.

Little, M. A. & Morren, G. E. B. (1977). *Ecology, Energetics and Human Variability*. Duburque, New York: William Brown.

Little, M. A., Galvin, K. & Leslie, P. W. (1988). Health and energy requirements of nomadic Turkana pastoralists. In *Coping with Uncertainty in Food Supply*. ed. I. de Garine & G. A. Harrison, pp. 290–317. Oxford: Oxford University Press.

Livesey, G. (1990). The energy values of unavailable carbohydrates and diets: an enquiry and analysis. *American Journal of Clinical Nutrition*, **51**, 617–37.

Livingstone, M. B. E., Stain, J. J., McKenna, P. G., Nevin, G. B., Barker, M. E., Hickey, R. J., Prentice, A. M., Coward, W. A., Ceesay, S. M. & Whitehead, R. G. (1990). Simultaneous measurement of free living energy expenditure by the double-labelled water ($^2H_2^{18}O$) method and heart rate monitoring. *American Journal of Clinical Nutrition*, **52**, 59–65.

van Loan, M. D. & Mayclin, P. L. (1992). Use of multi-frequency bioelectrical impedance analysis for the estimation of extracellular fluid. *European Journal of Clinical Nutrition,* **46**, 117–24.

Lohman, T. G. (1988). Anthropometry and body composition. In *Anthropometric Standardisation Reference Manual.* ed. T. G. Lohman, A. F. Roche & R. Martorell, pp. 125–9. Champaign, IL: Human Kinetics Books.

Lohman, T. G., Roche, A. F. & Martorell, R. (1988). Anthropometric Standardisation Reference Manual. Champaign, IL: Human Kinetics Books.

Longhurst, R. & Payne, P. (1981). Seasonal aspects of nutrition. In *Seasonal Dimensions to Rural Poverty.* ed. R. Chambers, R. Longhurst & A. Pacey, pp. 45–52. London: Frances Pinter.

Loudon, A. (1987). Nutritional effects on puberty and lactational infertility in mammals: some interspecies considerations. *Proceedings of the Nutrition Society,* **46**, 203–16.

Loutan, L. & Lamotte, J. M. (1984). Seasonal variations in nutrition among a group of nomadic pastoralists in Niger. *Lancet,* **2**, 945–7.

Lovelady, C. A., Meredith, C. N., McCrory, M. A., Nommsen, L. A., Joseph, L. J. & Dewey, K. G. (1993). Energy expenditure in lactating women: a comparison of doubly labeled water and heart-rate-monitoring methods. *American Journal of Clinical Nutrition,* **57**, 512–18.

Lucas, A. (1990). Energy requirements in normal infants and children. In *Activity, Energy Expenditure and Energy Requirements of Infants and Children.* ed. B. Schurch & N. S. Scrimshaw, pp. 9–34. Lausanne: International Dietary Energy Consultancy Group.

Lukaski, H. C., Johnson, P. E., Bolonchuk, W. W. & Lykken, G. I. (1985). Assessment of fat-free mass using bioelectrical impedance measurements of the human body. *American Journal of Clinical Nutrition,* **41**, 810–7.

Lukaski, H. C., Bolonchuk, W. W., Hall, C. B. & Siders, W. A. (1986). Validation of tetrapolar bioelectrical impedance method to assess human body composition. *Journal of Applied Physiology,* **60**, 1327–32.

Lunn, P. G. (1985). Maternal nutrition and lactational infertility: the baby in the driving seat. In *Maternal nutrition and lactational infertility.* ed. J. Dobbing, pp. 41–64. New York: Raven Press.

Lunn, P. G. (1988). Malnutrition and fertility. In *Natural Human Fertility.* ed. P. Diggory, M. Potts & S. Teper, pp. 135–52. Basingstoke: Macmillan Press.

Lunn, P. G. (1992). Breast-feeding patterns, maternal milk output and lactational infecundity. *Journal of Biosocial Science,* **24**, 317–24.

Lunn, P. G. (1994). Lactation and other metabolic loads affecting human reproduction. In *Human Reproductive Ecology. Interactions of Environment, Fertility, and Behavior.* ed. K. L. Campbell & J. W. Wood, pp. 77–85. New York: New York Academy of Sciences.

Lunn, P. G., Austin, S., Prentice, A. M. & Whitehead, R. G. (1984). The effect of improved nutrition on plasma prolactin concentrations and postpartum infertility in lactating Gambian women. *American Journal of Clinical Nutrition,* **39**, 227–35.

Lutter, J. M. & Cushman, S. (1982). Menstrual patterns in female runners. *Sports Medicine,* **10**, 60–72.

Malcolm, L. A. (1970). *Growth and Development in New Guinea – a Study of the Bundi People of the Madang District.* Madang: Institute of Human Biology.

Maloiy, G. M. O., Heglund, N. C., Prager, L. M., Cavagna, G. A. & Taylor, C. R. (1986). Energetic cost of carrying loads: have African women discovered an economic way? *Nature*, **319**, 668–9.

Margetts, B. M. & Nelson, M. (eds.) (1991). *Design Concepts in Nutritional Epidemiology*. Oxford: Oxford University Press.

Martin, R. D. (1983). Human brain evolution in an ecological context. *The 52nd James Arthur Lecture on the Evolution of the Brain*. New York: American Museum of Natural History.

Martin, R. D. (1989). *Primate Origins and Evolution: A Phylogenetic Reconstruction*. London: Chapman and Hall.

Martorell, R. (1985). Child growth retardation: a discussion of its causes and its relationship to health. In *Nutritional Adaptation in Man*. ed. K. Blaxter & J. C. Waterlow, pp. 13–30. London: John Libbey.

Martorell, R. & Arroyave, G. (1988). Malnutrition, work output and energy needs. In *Capacity for Work in the Tropics*. ed. K. J. Collins & D. F. Roberts, pp. 57–75. Cambridge: Cambridge University Press.

Martorell, R. & Habicht, J. P. (1986). Growth in early childhood in developing countries. In *Human Growth: A Comprehensive Treatise*, Vol. 3. ed. F. Falkner & J. M. Tanner, pp. 241–62. New York: Plenum Press.

Martorell, R., Yarbrough, C., Klein, R. E. & Lechtig, A. (1979). Malnutrition, body size, and skeletal maturation: interrelationships and implications for catch-up growth. *Human Biology*, **51**, 371–89.

Martorell, R., Rivera, J. & Kaplowitz, H. (1990). Consequences of stunting in early childhood for adult body size in rural Guatemala. *Nestlé Foundation Annual Report*, pp. 85–92. Lausanne: Nestlé Foundation.

Martorell, R., Rivera, J., Kaplowitz, H. & Pollitt, E. (1992). Long-term consequences of growth retardation during early childhood. In *Human Growth: Basal and Clinical Aspects*. ed. M. Hernandez & J. Argente, pp. 143–9. Amsterdam: Elsevier Science.

Martorell, R., Khan, L. K. & Schroeder, D. G. (1994). Reversibility of stunting: epidemiological findings in children from developing countries. *European Journal of Clinical Nutrition*, **48**(Supplement 1), S45–57.

Mason, E. D. & Jacob, M. (1972). Variations in basal metabolic rate responses to changes between tropical and temperate climates. *Human Biology*, **44**, 141–72.

Mata, L. J. (1983). Epidemiology of acute diarrhoea in childhood. In *Acute Diarrhoea: its Nutritional Consequence in Children*, ed. J. A. Bellanti, pp. 3–22. New York: Raven Press.

Mayes, P. A. (1988). Oxidative phosphorylation and mitochondrial transport systems. In *Harper's Biochemistry*, 21st edn. ed. R. K. Murray, D. K. Granmer, P. A. Mayes & V. W. Rodwell, pp. 108–19. Norwalk, CT: Appleton & Lange.

Mazess, R. B., Peppler, W. W., Chestnut, C. H., Nelp, W. B., Cohn, S. H. & Zanzi, I. (1981). Total body bone mineral and lean mass by dual-photon absorptiometry. II. Comparison with total body calcium by neutron activation analysis. *Calcified Tissue International*, **33**, 361–3.

Mazess, R. B., Peppler, W. W. & Gibbons, M. (1984). Total body composition by dual-photon (153 Gd) absorptiometry. *American Journal of Clinical Nutrition*, **40**, 834–9.

Mazess, R. B., Burden, H. S., Bisek, J. P. & Hanson, J. (1990). Dual-energy X-ray absorptiometry for total-body and regional bone-mineral and soft-tissue composition. *American Journal of Clinical Nutrition*, **51**, 1106–12.

McArthur, A. J. & Clark, J. A. (1987). Body temperature and heat and water balance. *Nature*, **326**, 647–8.

McArthur, M. (1977). Nutritional research in Melanesia: a second look at the Tsembaga. In *Subsistence and Survival: Rural Ecology in the Pacific*. ed. T. Bayliss-Smith & R. G. Feachem, pp. 91–128. London: Academic Press.

McCance, R. A. & Lawrence, R. D. (1929). The carbohydrate content of foods. *Medical Research Council Special Report No. 135*. London: Her Majesty's Stationary Office.

McGregor, I. A., Billewicz, W. Z. & Thomson, A. M. (1961), Growth and mortality in children in an African village. *British Medical Journal*, **2**, 1661–6.

McGuire, J. S. & Torun, B. (1984). Dietary energy intake and energy expenditure of women in rural Guatemala. In *Protein-Energy-Requirement Studies in Developing Countries: Results of International Research*. ed. W. Rand, R. Uauy & N. S. Scrimshaw, pp. 175–86. Tokyo: United Nations University.

McLean, J. A. & Tobin, G. (1987). *Animal and Human Calorimetry*. Cambridge: Cambridge University Press.

McNeill, G. & Payne, P. R. (1985). Energy expenditure of pregnant and lactating women. *Lancet* **i**, 1237–8.

McNeill, G., Rivers, J. P. W., Payne, P. R., de Britto, J. J. & Abel, R. (1987). Basal metabolic rate of Indian men: no evidence of metabolic adaptation to a low plane of nutrition. *Human Nutrition: Clinical Nutrition*, **41C**, 473–84.

McNeill, G., Payne, P. R., Rivers, J. P. W., Enos, A. M. T., De Britto, J. & Mukarji, D. S. (1988). Socio-economic and seasonal patterns of adult energy nutrition in a south Indian village. *Ecology of Food Nutrition*, **22**, 85–95.

McNeilly, A. S. (1988). Suckling and the control of gonadotropin secretion. In *The Physiology of Reproduction*. ed. E. Knobil & J. Neill, pp. 2323–49. New York: Raven Press.

McNeilly, A. S. (1994). Suckling and the control of gonadotropin secretion. In *The Physiology of Reproduction*. ed. E. Knobil & J. Neill. New York: Raven Press, in press.

McNeilly, A. S., Glasier, A. & Howie, P. W. (1985). Endocrine control of lactational infertility. In *Maternal Nutrition and Lactational Infertility*. ed. J. Dobbing, pp. 1–16. New York: Raven Press.

McNeilly, A. S., Howie, P. W. & Glasier, A. (1988). Lactation and the return of ovulation. In *Natural Human Fertility. Social and Biological Determinants*. ed. P. Diggory, M. Potts & S. Teper, pp. 102–17. London: MacMillan Press.

McNeilly, A. S., Tay, C. C. K. & Glasier, A. (1994). Physiological mechanisms underlying lactational amenorrhea. In *Human Reproductive Ecology. Interactions of Environment, Fertility, and Behavior*. ed. K. L. Campbell & J. W. Wood, pp. 145–55. New York: New York Academy of Sciences.

Meijer, G. A. L., Westerterp, K. R., Saris, W. H. M. & ten Hoor, F. (1992). Sleeping metabolic rate in relation to body composition and the menstrual cycle. *American Journal of Clinical Nutrition*, **55**, 637–40.

Messer, E. (1988). Seasonal hunger and coping strategies: an anthropological discussion. In *Coping with Seasonal Constraints*. ed. R. A. Huss-Ashmore, J. J.

Curry & R. K. Hitchcock, pp. 131–41. Philadelphia: MASCA Research Papers in Science and Archaeology.

Messer, E. (1989). Seasonality in food systems: an anthropological perspective on household food security. In *Seasonal Variability in Third World Agriculture. The Consequences for Food Security.* ed. D. E. Sahn, pp. 151–75. Baltimore: Johns Hopkins University Press.

Millar, J. G. (1965). Living systems: basic concepts. *Behavioural Science*, **10**, 193–237.

Millward, D. J. (1986). Hormonal responses to low intakes in relation to adaptation. In *Proceedings of the XIII International Congress of Nutrition.* ed. T. G. Taylor & N. K. Jenkins, pp. 419–23. London: John Libbey.

Milton, K. (1991). Comparative aspects of diet in Amazonian forest-dwellers. *Philosophical Transactions of the Royal Society of London, Series B*, **334**, 253–63.

Minghelli, G., Schutz, Y., Charbonnier, A., Whitehead, R. & Jequier, E. (1990). Reduced 24 hour energy expenditure and basal metabolic rate in Gambian men. *American Journal of Clinical Nutrition*, **51**, 563–70.

Minghelli, G., Schutz, Y., Whitehead, R. & Jequier, E. (1991). Seasonal changes in 24-h and basal energy expenditures in rural Gambian men as measured in a respiration chamber. *American Journal of Clinical Nutrition*, **53**, 14–20.

Ministry of Agriculture. (1973). *Farm Incomes in England and Wales, 1971–2.* London: Her Majesty's Stationery Office.

Molla, A. M., Molla, A., Sarker, S. A. & Rahaman, M. (1982). Food intake during and after recovery from diarrhoea in children. In *Diarrhoea and Malnutrition.* ed. L. C. Chen & N. S. Scrimshaw, pp. 113–23. New York: Plenum Press.

Molla, A. M., Molla, A. and Khatun, N. (1986). Absorption of macronutrients in children during acute diarrhoea and after recovery. In *Proceedings of the XIII International Congress of Nutrition.* ed. T. G. Taylor and N. K. Jenkins, pp. 113–15. London: John Libbey.

Montgomery, E. & Johnson, A. (1977). Machiguenga energy expenditure. *Ecology of Food and Nutrition*, **6**, 97–105.

Moran, E. F. (1982). *Human Adaptability. An Introduction to Ecological Anthropology.* Boulder, CO: Westview Press.

Moran, E. F. (1990). Ecosystem ecology in biology and anthropology: a critical assessment. In *The Ecosystem Approach in Anthropology.* ed. E. F. Moran, pp. 3–40. Ann Arbor: University of Michigan Press.

Morgan, R. W., Jain, M., Miller, A. B. *et al.* (1978). A comparison of dietary methods in epidemiological studies. *American Journal of Epidemiology*, **107**, 488–98.

Morphy, H. (1993). Cultural adaptation, In *Human Adaptation.* ed. G. A. Harrison, pp. 99–105. Oxford: Oxford University Press.

Morren, G. E. B. (1977). From hunting to herding: pigs and the control of energy in montane New Guinea. In *Subsistence and Survival. Rural Ecology in the Pacific.* ed. T. P. Bayliss-Smith & R. G. Feacham, pp. 273–315. New York: Academic Press.

Moscovskaia Oblast (1967). *Moskovskaia Oblast' za 50 Let Statischeskiy Sbornik.* Moscow: Statistika.

Mueller, W. H. & Martorell, R. (1988). Reliability and accuracy of measurement.

In *Anthropometric Standardisation Manual.* ed. T. G. Lohman, A. F. Roche & R. Martorell, pp. 83–6. Champaign, IL: Human Kinetics Books.

Mulligan, K. & Butterfield, G. E. (1990). Discrepancies between energy intake and expenditure in physically active women. *British Journal of Nutrition,* **64**, 23–6.

Nabarro, D., Howard, P., Cassels, C., Pant, M., Wijga, A. & Padfield, N. (1988). The importance of infections and environmental factors as possible determinants of growth retardation in children. In *Linear Growth Retardation in Less Developed Countries.* ed. J. C. Waterlow, pp. 165–79. New York: Raven Press.

Nagy, K. A. (1990). Introduction. In *The Doubly-Labelled Water Method for Measuring Energy Expenditure. Technical Recommendations for use in Humans.* ed. A. M. Prentice, pp. 1–16. Vienna: International Atomic Energy Agency.

Nagy, L. E. & King, J. C. (1983). Energy expenditure of pregnant women at rest or walking self-paced. *American Journal of Clinical Nutrition,* **38**, 369–76.

National Academy of Sciences (1978). *Post-Harvest Losses in Developing Countries.* Washington DC: Report of the National Academy of Sciences.

National Center for Health Statistics (1977). NCHS Growth Curves for Children. *Publication No. (PHS) 78–1650.* Hyattsville, MD: United States Department of Health, Education and Welfare.

Nelms, J. D. (1982). Measurement of work and effort. Physiological aspects. In *Energy and Effort.* ed. G. A. Harrison, pp. 1–25. London: Taylor and Francis.

Nelson, M. (1993). Social-class trends in British diet, 1860–1980. In *Food, Diet and Economic Change Past and Present.* ed. C. Geissler & D. J. Oddy, pp. 101–20. Leicester: Leicester University Press.

Nelson, M., Black, A. E., Morris, J. A. & Cole, T. J. (1989). Between- and within-subject variation in nutrient intake from infancy to old age: estimating the number of days required to rank dietary intakes with desired precision. *American Journal of Clinical Nutrition,* **50**, 155–67.

Neumann, C. G. & Harrison, G. G. (1994). Onset and evolution of stunting in infants and children. Examples from the Human Nutrition Collaborative Research Support Program. Kenya and Egypt studies. *European Journal of Clinical Nutrition,* **48**(Supplement 1), S90–102.

Nietschmann, B. (1973). *Between Land and Water.* New York: Seminar Press.

Norgan, N. G. (1982). Human energy stores. In *Energy and Effort.* ed. G. A. Harrison, pp. 139–58. London: Taylor & Francis.

Norgan, N. G. (1990). Thermogenesis above maintenance in humans. *Proceedings of the Nutrition Society,* **49**, 217–26.

Norgan, N. G. (1995a). Field methods for assessing body composition. In *Human Body Composition.* ed. P. S. W. Davies & T. J. Cole. Cambridge: Cambridge University Press, in press.

Norgan, N. G. (1995b). Measurement and interpretation issues in laboratory and field studies of energy expenditure. *American Journal of Human Biology,* in press.

Norgan, N. G. & Durnin, J. V. G. A. (1980). The effects of six weeks of overfeeding on the body weight, body composition and energy metabolism of young men. *American Journal of Clinical Nutrition,* **33**, 978–88.

Norgan, N. G. & Ferro-Luzzi, A. (1978). Nutrition, physical activity and physical fitness in contrasting environments. In: *Nutrition, Physical Fitness, and Health.*

ed. J. Parizkova & V. A. Rogozkin, pp. 167–93. Baltimore: University Park Press.

Norgan, N. G., Ferro-Luzzi, A. & Durnin, J. V. G. A. (1974). The energy and nutrient intake and the energy expenditure of 204 New Guinea adults. *Philosophical Transactions of the Royal Society of London, Series B,* **268,** 309–48.

Norgan, N. G., Durnin, J. V. G. A. & Ferro-Luzzi, A. (1979). The composition of some New Guinea foods. *Papua New Guinea Agricultural Journal,* **30,** 25–39.

Norgan, N. G., Ferro-Luzzi, A. & Durnin, J. V. G. A. (1982). The body composition of New Guinean adults in contrasting environments. *Annals of Human Biology,* **9,** 343–53.

Norgan, N. G., Shetty, P., Baskaran, T. & Ferro-Luzzi, A. (1989). Seasonality in a Karnataka (South India) agricultural cycle: background and body weight changes. In *Seasonality in Agriculture. Its Nutritional and Productivity Implications. International Food Policy Research Institute Workshop,* Washington, January 26–7.

Nydon, J. & Thomas, R. B. (1989). Methodological procedures for analysing energy expenditure. In *Research Methods in Nutritional Anthropology.* ed. G. H. Pelto, P. J. Pelto & E. Messer, pp. 57–81. Tokyo: United Nations University.

Nyerges, A. E. (1988). Seasonal constraints in the Guinea savanna: Susu ecology in Sierra Leone. In *Coping with Seasonal Constraints.* ed. R. Huss-Ashmore, J. J. Curry & R. K. Hitchcock, pp. 86–95. Philadelphia: MASCA Research Papers in Science and Archaeology.

O'Connell, J. F. & Hawkes, K. (1981). Alyawara plant use and optimal foraging theory. In *Hunter–Gatherer Foraging Strategies. Ethnographic and Archaeological Analyses.* ed. B. Winterhalder & E. A. Smith, pp. 99–125. Chicago: University of Chicago Press.

O'Donnell, L., O'Meara, N., Owens, D., Johnson, A., Collins, P. & Tomkin, G. (1987). Plasma catecholamines and lipoproteins in chronic psychological stress. *Journal of the Royal Society of Medicine,* **80,** 339–42.

Odum, E. P. (1971). *Environment, Power and Society.* New York: Wiley Interscience.

Offringa, P. J. & Boersma, E. R. (1987). Will food supplementation in pregnant women decrease neonatal morbidity? *Human Nutrition: Clinical Nutrition,* **41C,** 311–15.

Ohtsuka, R. (1983). *Oriomo Papuans. Ecology of Sago-eaters in Lowland Papua.* Tokyo: University of Tokyo Press.

Ohtsuka, R. (1990). Food consumption and major nutrient intake. In *Population Ecology of Human Survival.* ed. R. Ohtsuka & T. Suzuki, pp. 91–100. Tokyo: University of Tokyo Press.

Oleshansky, M. & Meyerhoff, J. (1992). Acute catecholaminergic responses to mental and physical stressors in man. *Stress Medicine,* **8,** 175–9.

Ordy, J. M., Samorajski, T., Zimmerman, R. R. & Rady, P. M. (1966). Effects of postnatal protein deficiency on weight gain, serum proteins, enzymes, cholesterol, and liver ultrastructure in a subhuman primate (*Macaca mulatta*). *American Journal of Pathology,* **48,** 769–91.

Orr-Ewing, A. K. (1985). A longitudinal study of infant growth in relation to breast milk consumption, total food intake and morbidity. Ph.D. thesis: University of London.

Orr-Ewing, A. K., Heywood, P. F. & Coward, W. A. (1986). Longitudinal measurements of breast milk output by a ^2H_2O tracer technique in rural Papua New Guinea women. *Human Nutrition: Clinical Nutrition*, **40C**, 451–67.

Orubuloye, I. O. (1981). Child-spacing among rural Yoruba women: Ekiti and Ibadan divisions in Nigeria. In *Child-Spacing in Tropical Africa: Traditions and Change*. ed. H. J. Page & R. Lesthaeghe, pp. 225–36. New York: Academic Press.

Ou, L. C. & Tenney, S. M. (1970). Properties of mitochondria from hearts of cattle acclimated to high altitude. *Respiratory Physiology*, **8**, 151–9.

Owen, O. E., Holup, J. L., d'Allesio, D. A. *et al.* (1987). A reappraisal of the caloric requirements of men. *American Journal of Clinical Nutrition*, **46**, 875–85.

Pacey, A. & Payne, P. (1985). *Agricultural Development and Nutrition*. London: Hutchinson.

Pagezy, H. (1982). Seasonal hunger, as experienced by the Oto and the Twa of a Ntommba village in the equatorial forest (Lake Tumba, Zaire). *Ecology of Food and Nutrition*, **12**, 139–53.

Palmon, A., Aroya, N. B., Tel-Or, S., Burstein, Y., Fridkin, M. & Koch, Y. (1994). The gene for the neuropeptide gonadotrophin-releasing hormone is expressed in the mammary gland of the rat. *Proceedings of the National Academy of Sciences USA*, **91**, 4994–6.

Panter-Brick, C. (1989). Motherhood and subsistence work: the Tamang of rural Nepal. *Human Ecology*, **17**, 205–28.

Panter-Brick, C. (1993). Seasonal organisation of work patterns. In *Seasonality and Human Ecology*. ed. S. J. Ulijaszek & S. S. Strickland, pp. 220–34. Cambridge: Cambridge University Press.

Panter-Brick, C., Lotstein, D. S. & Ellison, P. T. (1993). Seasonality of reproductive function and weight loss in rural Nepali women. *Human Reproduction*, **8**, 684–90.

Park, R. E. (1936). Human ecology. *American Journal of Sociology*, **42**, 1–15.

Parker, S. T. & Gibson, K. R. (1979). A model of the evolution of intelligence and language in early hominids. *Behavioural and Brain Sciences*, **2**, 367–407.

Pasquet, P. & Koppert, G. J. A. (1993). Activity patterns and energy expenditure in Cameroonian tropical forest populations. In *Tropical Forests, People and Food. Biocultural Interactions and Applications to Development*. ed. C. M. Hladik, A. Hladik, O. F. Linares, H. Pagezy, A. Semple & M. Hadley, pp. 311–20. Paris: UNESCO Publications.

Passmore, R. & Durnin, J. V. G. A. (1955). Human energy expenditure. *Physiology Reviews*, **35**, 801–40.

Paul, A. A. & Southgate, D. A. T. (1978). *The Composition of Foods*. London: Her Majesty's Stationery Office.

Payne, P. R. (1989). Public health and functional consequences of seasonal hunger and malnutrition. In *Seasonal Variability in Third World Agriculture. The Consequences for Food Security*. ed. D. E. Sahn, pp. 19–46. Baltimore: Johns Hopkins University Press.

Payne, P. R. & Dugdale, A. E. (1977). A model for the prediction of energy balance and body weight. *Annals of Human Biology*, **4**, 525–35.

Pelletier, D. L. (1991). *Relationships Between Child Anthropometry and Mortality in Developing Countries: Implications for Policy, Programs, and Future Research.* Ithaca, NY: Cornell Food and Nutrition Policy Program Monograph No. 12.

Pennington, R. & Harpending, H. (1988): Fitness and fertility among Kalahari !Kung. *American Journal of Physical Anthropology*, **77**, 303–19.

Perkins, J. F. (1954). Plastic Douglas bags. *Journal of Applied Physiology*, **6**, 445–7.

Pianka, E. R. (1988). *Evolutionary Ecology*, 4th edn. New York: Harper & Row.

Piers, L. S. & Shetty, P. S. (1993). Basal metabolic rates of Indian women. *European Journal of Clinical Nutrition*, **47**, 586–91.

Pietinen, P., Hartmen, A. M., Haapa, E. *et al.* (1988). Reproducibility and validity of dietary assessment instruments. I. A self-administered food use questionnaire with a portion size picture booklet. *American Journal of Epidemiology*, **128**, 655–66.

Pimentel, D. & Pimentel, M. (1979). *Food, Energy and Society*. London: Edward Arnold.

Pollack, A. & Steklis, H. (1986). Urinary catecholamines and stress in male and female police cadets. *Human Biology*, **58**, 209–20.

Poppitt, S. D., Prentice, A. M., Jequier, E., Schutz, Y. & Whitehead, R. G. (1993). Evidence of energy sparing in Gambian women during pregnancy: a longitudinal study using whole-body calorimetry. *American Journal of Clinical Nutrition*, **57**, 353–64.

Prader, A., Tanner, J. M. & von Harnack, G. A. (1963). Catch-up growth following illness or starvation. *Journal of Paediatrics*, **62**, 645–59.

Prentice, Ann (1986). The effect of maternal parity on lactational performance in a rural African community. In *Human Lactation 2*. ed. M. Hamosh & A. S. Goldman, pp. 165–73. New York: Plenum Press.

Prentice, Ann & Bates, C. J. (1994). Adequacy of dietary mineral supply for human bone growth and mineralisation. *European Journal of Clinical Nutrition*, **48**(Supplement 1), S161–76.

Prentice, Ann & Paul, A. (1990). Contribution of breast-milk to nutrition during prolonged breast-feeding. In *Human Lactation 4: Breast-Feeding, Nutrition, Infections and Infant Growth*. ed. S. Atkinson, L. Hanson & R. Chandra, pp. 87–102. St Johns, Newfoundland: ARTS Biomedical.

Prentice, Ann, Prentice, A. M., Cole, T. J., Paul, A. A. & Whitehead, R. G. (1984). Breast-milk antimocrobial factors of rural Gambian mothers. 1. Influence of stage of lactation and maternal plane of nutrition. *Acta Paediatrica Scandinavica*, **73**, 796–802.

Prentice, Ann, Cole, T. J. & Whitehead, R. G. (1987a). Impaired growth in infants born to mothers of very high parity. *Human Nutrition: Clinical Nutrition*, **41C**, 319–25.

Prentice, A. M. (1980). Variations in maternal dietary intake, birthweight and breastmilk output in The Gambia. In *Maternal Nutrition During Pregnancy and Lactation*. ed. H. Aebi & R. G. Whitehead, pp. 167–83. Bern: Hans Huber.

Prentice, A. M. (1988). Applications of the doubly-labelled water ($^2H_2$, ^{18}O) method in free-living adults. *Proceedings of the Nutrition Society*, **47**, 259–68.

Prentice, A. M. (1990a). Long-term energy expenditure measurement: stable isotope method. In *Handbook of Methods for the Measurement of Work Performance, Physical Fitness and Energy Expenditure in Tropical Populations.*

ed. K. J. Collins, pp. 91–4. Paris: International Union of Biological Sciences.

Prentice, A. M. (1990b). *The Doubly-Labelled Water Method for Measuring Energy Expenditure. Technical Recommendations for use in Humans.* Vienna: International Atomic Energy Agency.

Prentice, A. M. & Prentice, A. (1990). Maternal energy requirements to support lactation. In *Breastfeeding, Nutrition, Infection and Infant Growth in Developed and Emerging Countries.* ed. S. A. Atkinson, L. A. Hanson & R. K. Chandra, pp. 67–86. St John's, Newfoundland: Biomedical Publishers.

Prentice, A. M., Whitehead, R. G., Roberts, S. B. *et al.* (1980). Dietary supplementation of Gambian nursing mother and lactational performance. *Lancet* i, 886–8.

Prentice, A. M., Whitehead, R. G., Roberts, S. B. & Paul, A. A. (1981). Long-term energy balance in child-bearing Gambian women. *American Journal of Clinical Nutrition*, **34**, 2790–9.

Prentice, A. M., Whitehead, R. G., Watkinson, M., Lamb, W. H. & Cole, T. J. (1983). Prenatal dietary supplementation of African women and birthweight. *Lancet* i, 489–92.

Prentice, A. M., Paul, A. A., Prentice, A., Black, A., Cole, T. J. & Whitehead, R. G. (1986a). Cross-cultural differences in lactational performance. In *Human Lactation 2.* ed. M. Hamosh and A. S. Goldman. New York: Plenum Press.

Prentice, A. M., Black, A. E., Coward, W. A., Davies, H. L., Goldberg, G. R., Murgatroyd, P. R., Ashford, J., Sawyer, M. & Whitehead, R. G. (1986b). High levels of energy expenditure in obese women. *British Medical Journal*, **292**, 983–7.

Prentice, A. M., Cole, T. J., Foord, F. A., Lamb, W. H. & Whitehead, R. G. (1987b). Increased birthweight after prenatal dietary supplementation of rural African women. *American Journal of Clinical Nutrition*, **46**, 912–25.

Prentice, A. M., Goldberg, G. R., Davies, H. L., Murgatroyd, P. R. & Scott, W. (1989). Energy sparing adaptations in human pregnancy assessed by whole-body calorimetry. *British Journal of Nutrition*, **62**, 5–22.

Prentice, A. M., Vasquez-Velasquez, L., Davies, P. S. W., Lucas, A. & Coward, W. A. (1990). Total energy expenditure of free-living infants and children obtained by the doubly-labelled water method. In *Activity, Energy Expenditure and Energy Requirements of Infants and Children.* ed. B. Schurch & N. S. Scrimshaw, pp. 83–101. Lausanne: International Dietary Energy Expenditure Group.

Pullicino, E., Coward, W. A., Stubbs, R. J. & Elia, M. (1990). Bedside and field methods for assessing body composition: comparison with the deuterium dilution technique. *European Journal of Clinical Nutrition*, **44**, 753–62.

Pyke, G. H., Pulliam, H. R. & Charnov, E. L. (1977). Optimal foraging: A selective review of theory and tests. *Quarterly Review of Biology*, **52**, 137–54.

Quenouille, M. H., Boyne, A. W., Fisher, W. B. & Leitch, I. (1951). Statistical studies of recorded energy expenditure of man. Part I. Basal metabolism related to sex, stature, age, climate and race. *Technical Communication No. 17.* Bucksburn, Aberdeenshire: Commonwealth Bureau of Animal Nutrition.

Quigley, M. E. & Yen, S. S. C. (1980). The role of endogenous opiates on luteinising hormone secretion during the menstrual cycle. *Journal of Clinical Endocrinology and Metabolism*, **51**, 179–81.

van Raaij, J. M. A., Schonk, C. M., Vermaat-Miedema, S. H., Peek, M. E. M. & Hautvast, J. G. A. J. (1989). Body fat mass and basal metabolic rate in Dutch women before, during, and after pregnancy: a reappraisal of energy cost of pregnancy. *American Journal of Clinical Nutrition*, **49**, 765–72.

van Raaij, J. M. A., Schonk, C. M., Vermaat-Miedema, S. H., Peek, M. E. M. & Hautvast, J. G. A. J. (1990). Energy cost of physical activity throughout pregnancy and the first year postpartum in Dutch women with sedentary lifestyles. *American Journal of Clinical Nutrition*, **52**, 234–9.

van Raaij, J. M. A., Schonk, C. M., Vermaat-Miedema, S. H., Peek, M. E. M. & Hautvast, J. G. A. J. (1991). Energy cost of lactation, and energy balances of well-nourished Dutch lactating women: reappraisal of the extra energy requirements of lactation. *American Journal of Clinical Nutrition*, **53**, 612–19.

Rahaman, M. M., Yamauchi, M., Hanada, N., Nishikawa, K. & Morishima, T. (1987). Local production of rotavirus specific IgA in breast tissue and transfer to neonates. *Archives of Disease in Childhood*, **62**, 401–5.

Rappaport, R. A. (1968). *Pigs of the Ancestors: Ritual in the Ecology of a New Guinea People*. New Haven: Yale University Press.

Rappaport, R. A. (1971). Ritual, sanctity and cybernetics. *American Anthropologist*, **73**, 59–76.

Rappaport, R. A. (1978). Adaptation and the structure of ritual. In *Human Behaviour and Adaptation*. ed. N. Blurton Jones & V. Reynolds, pp. 77–102. London: Taylor & Francis.

Reardon, T. (1989). Seasonality in household transactions in western Niger. Paper presented at the IFPRI workshop *Seasonality in agriculture. Its nutritional and productivity implications*. Washington, January 26–27.

Rebello, N. S. P., Chandrashekar, G. S., Shankaramurthy, H. G. & Hiramath, K. C. (1976). Impact of the increase in prices of inputs in Mandya District of Karnataka. *Indian Journal of Agricultural Economics*, **31**, 71–81.

Reddy, V. (1991). Body mass index and mortality rates. *Nutrition News*, Vol. 12. Hyderabad: National Institute of Nutrition.

Rising, R., Swinburn, B., Larson, K. & Ravussin, E. (1991). Body composition in Pima Indians: validation of bioelectrical resistance. *American Journal of Clinical Nutrition*, **53**, 594–8.

Roberts, S. B., Paul, A. A., Cole, T. J. & Whitehead, R. G. (1982). Seasonal changes in activity, birth weight and lactational performance in rural Gambian women. *Transactions of the Royal Society of Tropical Medicine and Hygiene*, **76**, 668–78.

Roberts, S. B., Coward, W. A., Ewing, G., Savage, J., Cole, T. J. & Lucas, A. (1988). Effect of weaning on accuracy of doubly-labelled water method in infants. *American Journal of Physiology*, **254**, R622–7.

Rode, A. & Shephard, R. J. (1994). Prediction of body fat content in an Inuit community. *American Journal of Human Biology*, **6**, 249–54.

Rodman, P. S. & McHenry, H. M. (1980). Bioenergetics and origins of bipedalism. *American Journal of Physical Anthropology*, **52**, 103–6.

Roede, M. J. & van Wieringen, J. C. (1985). Growth diagrams, 1980. *Tijdshrift voor Sociale Gezondheidszorg*, **63**, 1–34.

Rohr, H. P. & Riede, U. N. (1973). Experimental metabolic disorders and the subcellular reaction pattern. *Current Topics in Pathology*, **58**, 1–48.

Rosetta, L. (1986). Sex differences in seasonal variations of the nutritional status of Serere adults in Senegal. *Ecology of Food and Nutrition*, **18**, 231–44.

Rosetta, L. (1988a). Seasonal variations in food consumption by Serere families in Senegal. *Ecology of Food and Nutrition*, **20**, 275–86.

Rosetta, L. (1988b). Seasonal changes and the physical development of young Serere children in Senegal. *Annals of Human Biology*, **15**, 179–89.

Rosetta, L. (1990). Biological aspects of fertility among third world populations. In *Fertility and Resources*, ed. V. Reynolds & J. Landers, pp. 18–34. Cambridge: Cambridge University Press.

Rosetta, L. (1992). Aetiological approach of female reproductive physiology in lactational amenorrhoea. *Journal of Biosocial Science*, **24**, 301–15.

Rosetta, L. (1993). Female reproductive dysfunction and intense physical training. *Oxford Reviews of Reproductive Biology*, **15**, 113–41.

Rosetta, L. (1994). Nutrition, physical workloads and fertility. In *Reproductive Decisions: Biological and Social Perspectives*, ed. R. I. M. Dunbar. London: Macmillan Press, in press.

Rosso, P. (1985). A new chart to monitor weight gain during pregnancy. *American Journal of Clinical Nutrition*, **41**, 644–52.

Rosso, P. (1990). *Nutrition and Metabolism in Pregnancy*. Oxford: Oxford University Press.

Roubenoff, R., Kehayias, J. J., Dawson-Hughes, B. & Heymesfiled, S. B. (1993). Use of dual-energy X-ray absorptiometry in body-composition studies: not yet a 'gold standard'. *American Journal of Clinical Nutrition*, **58**, 589–91.

Rowland, M. G. M. & Rowland, S. G. J. G. (1986). Growth faltering in diarrhoea. In *Proceedings of the XIII International Congress of Nutrition*. ed. T. G. Taylor & N. K. Jenkins, pp. 115–19. London: John Libbey.

Rowland, M. G. M., Cole, T. J. & Whitehead, R. G. (1977). A quantitative study into the role of infection in determining nutritional status in Gambian village children. *British Journal of Nutrition*, **37**, 441–50.

Rowland, M. G. M., Paul, A., Prentice, A. M. *et al.* (1981). Seasonality and the growth of infants in a Gambian village. In *Seasonal dimensions to rural poverty*. ed. R. Chambers, R. Longhurst & A. Pacey, pp. 164–75. London: Frances Pinter.

Rowland, M. G. M., Rowland, S. G. J. G. & Cole, T. J. (1988). Impact of infection on the growth of children from 0 to 2 years in an urban West African community. *American Journal of Clinical Nutrition*, **47**, 134–8.

Sadurskis, A., Kabir, N., Wager, J. & Forsum, E. (1988). Energy metabolism, body composition, and milk production in healthy Swedish women during lactation. *American Journal of Clinical Nutrition*, **48**, 44–9.

Saha, N., Tan, P. Y. & Banerjee, B. (1985). Energy balance study in Singapore medical students. *Annals of Nutrition and Metabolism*, **29**, 216–22.

Sanborn, C. F., Martin, B. J. & Wagner, W. W. (1982). Is athletic amenorrhea specific to runners? *American Journal of Obstetrics and Gynecology*, **143**, 859–61.

Satyanarayana, K., Naidu, N. A., Chaterjee, B. & Narasinga Rao, B. S. (1977). Body size and work output. *American Journal of Clinical Nutrition*, **30**, 322–5.

Schoeller, D. A. (1983). Energy expenditure from doubly labelled water: some fundamental considerations in humans. *American Journal of Clinical Nutri-*

tion, **38**, 999–1005.

Schoeller, D. A. (1984). Use of two-point sampling for the doubly-labelled water method. *Human Nutritional: Clinical Nutrition,* **38C**, 477–80.

Schoeller, D. A. & Fjeld, C. R. (1991). Human energy metabolism: what have we learned from the doubly labeled water method? *Annual Reviews in Nutrition,* **11**, 355–73.

Schoeller, D. A. & van Santen, E. (1982). Measurement of energy expenditure in humans by doubly labelled water method. *Journal of Applied Physiology,* **63**, 955–9.

Schoeller, D. A. & Taylor, P. B. (1987). Precision of the doubly labelled water method using the two point calculation. *Human Nutrition: Clinical Nutrition,* **41C**, 215–23.

Schoeller, D. A., van Santen, E., Peterson, D. W., Dietz, W., Jaspen, J. & Klein, P. D. (1980). Total body water measurement in humans with 180 and 2H labeled water. *American Journal of Clinical Nutrition,* **33**, 2686–93.

Schoener, T. W. (1971). Theory of feeding strategies. *Annual Review of Ecology and Systematics,* **2**, 369–404.

Schofield, S. (1974). Seasonal factors affecting nutrition in different age groups and especially pre-school children. *Journal of Development Studies,* **11**, 22–40.

Schofield, W. N. (1985). Predicting basal metabolic rate, new standards and review of previous work. *Human Nutrition: Clinical Nutrition,* **39C**(Supplement 1), 5–41.

Scholl, T., Hediger, M. L., Ances, I. G., Belsky, D. H. & Salmon, R. W. (1990). Weight gain during pregnancy in adolescence: predictivability of early weight gain. *Obstetrics and Gynecology,* **75**, 948–53.

Schultinck, W. J., Klaver, W., van Wijk, H., van Raaij, J. M. A. & Hautvast, J. G. A. J. (1990). Body weight changes and basal metabolic rates of rural Beninese women during seasons with different energy intakes. *European Journal of Clinical Nutrition,* **44**(Supplement 1) 31–40.

Schutz, Y., Lechtig, A. & Bradfield, R. (1980). Energy expenditures and food intakes of lactating women in Guatamala. *American Journal of Clinical Nutrition,* **33**, 892–902.

Seckler, D. (1982). Small but healthy: a basic hypothesis in the theory, measurement and policy of malnutrition. In *Newer Concepts in Nutrition and their Implications for Policy.* ed. P. V. Sukhatme, pp. 127–37. Pune: Maharashtra Association for the Cultivation of Science.

Segal, K. R., van Loan, M., Fitzgerald, P. I., Hodgdon, J. A. & van Ittallie, T. B. (1988). Lean body mass estimation by bioelectrical impedance analysis: a four-site cross-validation study. *American Journal of Clinical Nutrition,* **47**, 7–14.

Segal, K. R., Burastero, S., Chun, A., Coronel, P., Pierson, R. N. & Wang, J. (1991). Estimation of extracellular and total body water by multiple frequency bioelectrical-impedance measurement. *American Journal of Clinical Nutrition,* **54**, 26–9.

Sepulveda, J., Willett, W. & Munoz, A. (1988). Malnutrition and diarrhoea. A longitudinal study among urban Mexican children. *American Journal of Epidemiology,* **127**, 365–76.

Seymour-Smith, C. (1986). *Macmillan Dictionary of Anthropology.* London:

Macmillan Press.

Shangold, M. M. & Levine, H. S. (1982). The effect of marathon training on menstrual function. *American Journal of Obstetrics and Gynecology*, **143**, 862–9.

Shephard, R. J. (1991). *Body Composition in Biological Anthropology*. Cambridge: Cambridge University Press..

Shephard, R. J., Hatcher, J. & Rode, A. (1973). On the body composition of the Eskimo. *European Journal of Applied Physiology*, **32**, 3–15.

Shetty, P. S. (1984). Adaptive changes in basal metabolic rate and lean body mass in chronic undernutrition. *Human Nutrition: Clinical Nutrition*, **38C**, 443–52.

Shetty, P. S. (1993). Chronic undernutrition and metabolic adaptation. *Proceedings of the Nutrition Society*, **52**, 267–84.

Shetty, P. S. & Kurpad, A. V. (1990). Role of the sympathetic nervous system in adaptation to seasonal energy deficiency. *European Journal of Clinical Nutrition*, **44**(Supplement 1), 47–53.

Shetty, P. S. & Soares, M. J. (1988). Variability in basal metabolic rates of man. In *Comparative Nutrition*. ed. K. Blaxter & I. MacDonald, pp. 141–8. London: John Libbey.

Siega-Riz, A. & Adair, L. S. (1993). Biological determinants of pregnancy weight gain in a Filipino population. *American Journal of Clinical Nutrition*, **57**, 365–72.

Silberbauer, G. (1981). Hunter gatherers of the Kalahari. In *Omnivorous Primates*. ed. R. Harding & G. Teleki, pp. 472. New York: Columbia University Press.

Simondon, K. B., Bénéfice, E., Simondon, F., Delaunay, V. & Chahnazarian, A. (1993). Seasonal variation in nutritional status of adults and children in rural Senegal. In *Seasonality and Human Ecology*, ed. S. J. Ulijaszek & S. S. Strickland, pp. 166–83. Cambridge: Cambridge University Press.

Singh, J., Prentice, A. M., Diaz, E., Coward, A. W., Ashford, J., Sawyer, M. & Whitehead, R. G. (1989). Energy expenditure of Gambian women during peak agricultural activity measured by the doubly-labelled water method. *British Journal of Nutrition*, **62**, 315–29.

Siri, W. E. (1956). *Body Composition from Fluid Spaces and Density: An Analysis of Methods*. Berkeley, CA: University of California Radiation Laboratory Publication No. 3349.

Sloan, A. W. (1967). Estimation of body fat in young men. *Journal of Applied Physiology*, **23**, 311–15.

Small, R. & Ulijaszek, S. J. (1993). Stress, energy expenditure and menstrual dysfunction in Cambridge women. *American Journal of Physical Anthropology*, Supplement 16, 182.

Smith, E. A. (1979). Human adaptation and energetic efficiency. *Human Ecology*, **7**, 53–74.

Smith, E. A. (1983). Optimal foraging theory and hunter–gatherer societies. *Current Anthropology*, **24**, 625–51.

Smith, E. A. (1984). Anthropology, evolutionary ecology, and the explanatory limitations of the ecosystem concept. In *The Ecosystem Concept in Anthropology*, ed. E. F. Moran, pp. 51–86. Washington, DC: American Association for the Advancement of Science.

Soares, M. J., Francis, D. G. & Shetty, P. S. (1993). Predictive equations for basal metabolic rates of Indian males. *European Journal of Clinical Nutrition*, **47**, 389–94.

Solomon, S. J., Kurzer, M. S. & Calloway, D. H. (1982). Menstrual cycle and basal metabolic rate in women. *American Journal of Clinical Nutrition*, **36**, 611–16.

Southgate, D. A. T. (1993). Food composition tables. In *Human Nutrition and Dietetics*. ed. J. Garrow & W. P. T. James, pp. 264–72. London: Churchill Livingstone.

Soysa, P. (1987). Women and nutrition. *World Review of Nutrition and Dietetics*, **52**, 1–70.

Spencer, T. & Heywood, P. (1983). Seasonality, subsistence agriculture and nutrition in a lowlands community of Papua New Guinea. *Ecology of Food and Nutrition*, **13**, 221–9.

Sperof, L. & Redwine, D. B. (1979). Exercise and menstrual dysfunction. *Physiological Sports Medicine*, **8**, 42–52.

Spurr, G. B. (1984). Physical activity, nutritional status and physical work capacity in relation to agricultural productivity. In *Energy Intake and Activity*. ed. E. Pollitt & P. Amante, pp. 207–61. New York: Alan R. Liss.

Spurr, G. B. (1988a). Body size, physical work capacity, and productivity in hard work: is bigger better? In *Linear Growth Retardation in Less Developed Countries*. ed. J. C. Waterlow, pp. 215–39. New York: Raven Press.

Spurr, G. B. (1988b). Marginal malnutrition in childhood: implications for adult work capacity and productivity. In *Capacity for Work in the Tropics*. ed. K. J. Collins & D. F. Roberts, pp. 107–40. Cambridge: Cambridge University Press.

Spurr, G. B. & Riena, J. C. (1990). Estimation and validation of energy expenditure obtained by the minute-by-minute measurement of heart-rate. In *Activity, Energy Expenditure and Energy Requirements of Infants and Children*. ed. B. Schurch & N. S. Scrimshaw, pp. 57–69. Lausanne: International Dietary Energy Consultancy Group.

Spurr, G. B., Maksud, M. G. & Barac-Nieto, M. (1977). Energy expenditure, productivity and physical work capacity of sugar cane loaders. *American Journal of Clinical Nutrition*, **30**, 1740–6.

Spurr, G. B., Prentice, A. M., Murgatroyd, P. R., Goldberg, G. R., Reina, J. C. & Christman, N. T. (1988). Energy expenditure from minute-by-minute heart rate recording: comparison with indirect calorimetry. *American Journal of Clinical Nutrition*, **48**, 552–9.

Srikantia, S. G. (1985). Nutritional adaptation in man. *Proceedings of the Nutrition Society of India*, **31**, 1–16.

Steckel, R. H. (1987). Growth depression and recovery: the remarkable case of American slaves. *Annals of Human Biology*, **14**, 111–32.

Stein, T. P., Hoyt, R. W., Settle, R. G., O'Toole, M. & Hiller, W. D. B. (1987). Determination of energy expenditure during heavy exercise, normal daily activity and sleep using the doubly labelled water ($^2H_2{}^{18}O$) method. *American Journal of Clinical Nutrition*, **45**, 534–9.

Stenning, D. J. (1971). Household viability among the pastoral Fani. In *The Development Cycle in Domestic Groups*. ed. J. Goody, pp. 92–119. Cambridge: Cambridge University Press.

Steward, J. (1955). *The Theory of Culture Change*. Urbana: University of Illinois Press.

Stoeckel, J. & Chowdhury, K. M. A. (1972). Seasonal variation in births in rural East Pakistan. *Journal of Biosocial Science*, **4**, 107–16.

Strickland, S. S. (1986). Long term development of Kejaman subsistence: an ecological study. *Sarawak Museum Journal*, XXXVI, 117–71.

Strickland, S. S. (1990). Traditional economies and patterns of nutritional disease. In *Diet and Disease*. ed. G. A. Harrison & J. C. Waterlow, pp. 209–39. Cambridge: Cambridge University Press.

Strickland, S. S. & Ulijaszek, S. J. (1990). Energetic cost of standard activities in Gurkha and British soldiers. *Annals of Human Biology*, **17**, 133–44.

Strickland, S. S. & Ulijaszek, S. J. (1993). Body mass index, ageing and differential reported morbidity in rural Sarawak. *European Journal of Clinical Nutrition*, **47**, 9–19.

Strickland, S. S., Tuffrey, V. R., Gurung, G. M. & Ulijaszek, S. J. (1993). *Micro-Economic Consequences of Physique and Physical Performance in South Asia. Overseas Development Administration Report*. London: Overseas Development Administration.

Stuff, J. E., Garza, C., O'Brien-Smith, E., Nichols, B. L. & Montandon, C. M. (1983). A comparison of dietary methods in nutritional studies. *American Journal of Clinical Nutrition*, **37**, 300–6.

Svobada, D. & Higginson, J. (1964). Ultrastructural changes produced by protein and related deficiencies in the rat liver. *American Journal of Pathology*, **45**, 353–79.

Swift, J. (1981). Labour and subsistence in a pastoral economy. In *Seasonal Dimensions to Rural Poverty*. ed. R. Chambers, R. Longhurst & A. Pacey, pp. 80–7. London: Frances Pinter.

Takasaka, K. (1986). Postpartum amenorrhea, waiting time to conception, and prevalence of pregnancy of women in a Sundanese agricultural community. *Human Biology*, **58**, 933–44.

Tanaka, K., Inagaki, A., Matsuura, Y., Nakadomo, F., Hazama, T. & Maeda, K. (1990). Validity of objectivity of bioelectrical impedance technique for body composition assessment. *Clinical Sports Medicine*, **7**, 939–45.

Tanaka, K., Nakadomo, F., Watanabe, K., Inagaki, A., Kim, H. K. & Matsuura, Y. (1992). Body composition prediction equations based on bioelectrical impedance and anthropometric variables for Japanese obese women. *American Journal of Human Biology*, **4**, 739–45.

Tandler, B. & Hoppel, C. L. (1986). Studies on giant mitochondria. *Annals of the New York Academy of Sciences*, **488**, 65–81.

Tanner, J. M. (1989). *Foetus into Man. Physical Growth from Conception to Maturity*, 2nd edn. Ware: Castlemead Publications.

Tansley, A. G. (1946). *Introduction to Plant Ecology*. London: Allen and Unwin.

Tay, C. C. K., Glasier, A. F. & McNeilly, A. S. (1993). Effect of antagonists of dopamine and opiates on the basal and GnRH-induced secretion of luteinising hormone, follicle stimulating hormone and prolactin during lactational amenorrhoea in breastfeeding women. *Human Reproduction*, **8**, 532–9.

Tembon, A. C. (1990). Seasonality of births in the North West Province, Cameroon: implications for family planning programme. *Central African Journal of Medicine*, **36**, 90–3.

Terjung, R. L. & Kaciuba-Uscilko, H. (1986). Lipid metabolism during exercise:

influence of training. *Diabetes Metabolism Review*, **2**, 35–51.

Thomas, R. B. (1973a). Human adaptation to a high Andean energy-flow system. *Occasional Paper in Anthropology No. 7*. University Park: Pennsylvania State University Department of Anthropology.

Thomas, R. B. (1973b). The ecology of work. In *Physiological Anthropology*. ed. A. Damon, pp. 59–79. Oxford: Oxford University Press.

Thomas, R. B. (1976). Energy flow at high altitude. In *Man in the Andes: a Multidisciplinary Study of High-altitude Quechua*. ed. P. T. Baker & M. A. Little, pp. 379–404. Stroudsburg, PA: Dowden, Hutchinson and Ross.

Thomas, R. B. & Leatherman, T. L. (1990). Household coping strategies and contradictions in response to seasonal food shortage. *European Journal of Clinical Nutrition*, **44**(Supplement 1), 103–11.

Thomas, R. B., McRae, S. D. & Baker, P. T. (1982). The use of models in anticipating effects of change on human populations. In *Energy and Effort*. ed. G. A. Harrison, pp. 243–81. London: Taylor and Francis.

Thomas, R. B., Gage, T. B. & Little, M. A. (1989). Reflections on adaptive and ecological models. In *Human Population Biology*. ed. M. A. Little & J. D. Haas, pp. 296–319. Oxford: Oxford University Press.

Thomasset, A. (1963). Bio-electrical properties of tissue. *Lyon Medecine*, **209**, 1325–62.

Thomson, A. M. & Billewicz, W. Z. (1957). Clinical significance of weight trends during pregnancy. *British Medical Journal*, **1**, 243–7.

Tin-May-Than & Ba-Aye (1985). Energy intake and energy output of Burmese farmers at different seasons. *Human Nutrition: Clinical Nutrition*, **39C**, 7–15.

Tomkins, A. M. (1983). Nutritional cost of protracted diarrhoea in young Gambian children. *Gut*, **24**, A549.

Tomkins, A. M. (1986a). Protein energy malnutrition and risk of infection. *Proceedings of the Nutrition Society*, **45**, 289–304.

Tomkins, A. M. (1986b). Nutrient intake during diarrhoea in young children. In *Proceedings of the XIII International Congress of Nutrition*. ed. T. G. Taylor & N. K. Jenkins, pp. 110–13. London: John Libbey.

Tomkins, A. (1988). The risk of morbidity in a stunted child. In *Linear Growth Retardation in Less Developed Countries*. ed. J. C. Waterlow, pp. 185–95. New York: Raven Press.

Tomkins, A. M. & Watson, F. (1989). *Malnutrition and Infection. Clinical Nutrition Unit Report*. London: London School of Hygiene and Tropical Medicine.

Toriola, A. L. (1988). Survey of menstrual function in young Nigerian athletes. *International Journal of Sports Medicine*, **9**, 29–34.

Tripathi, A. M., Agarwal, K. N., Agarwal, D. K., Devi, R. R. & Cherian, S. (1987). Nutritional status of rural pregnant women and fetal outcome. *Indian Pediatrics*, **24**, 703–12.

Truswell, A. S. & Hansen, J. D. L. (1976). Medical research among the !Kung. In *Kalahari Hunter-Gatherers*. ed. R. B. Lee & I de Vore, pp. 166–94. Cambridge, MA: Harvard University Press.

Tuazon, M. A. G., van Raaij, J. M. A., Hautvast, J. G. A. G. & Barba, C. V. C. (1987). Energy requirements of pregnancy in the Philippines. *Lancet* ii, 1129–31.

Turnbull, C. M. (1974). Demography of small-scale societies. In *Dimensions of*

Society. ed. D. Polter & P. Sarre. pp. 9–41. Sevenoaks, UK: Hodder and Stoughton.

Ulijaszek, S. J. (1985). Seasonal variation of food intake in Ningerum villagers of Papua New Guinea. Poster presented at the *XIII International Congress of Nutrition*, Brighton, UK, 18–23 August.

Ulijaszek, S. J. (1987). Nutritional anthropometry, with special reference to populations in Papua New Guinea and Britain. PhD thesis. London: University of London.

Ulijaszek, S. J. (1990a). Comments on 'Steps toward an integrative medical anthropology'. *Medical Anthropology Quarterly*, 5, 374–9.

Ulijaszek, S. J. (1990b). Nutritional status and susceptibility to infectious disease. In *Diet and Disease in Traditional and Developing Societies*. ed. G. A. Harrison & J. C. Waterlow, pp. 137–54. Cambridge: Cambridge University Press.

Ulijaszek, S. J. (1991). Traditional methods of sago palm management in the Purari delta of Papua New Guinea. In *Proceedings of the 4th International Sago Symposium*, Kuching, Sarawak, Malaysia. ed. N. Thai-Tsiung, T. Yiu-Liong & K. Hong-Siong, pp. 122–6. Kuching, Sarawak: Ministry of Agriculture and Community Development, and Department of Agriculture.

Ulijaszek, S. J. (1992a). Human energetics methods in biological anthropology. *Yearbook of Physical Anthropology*, 35, 215–42.

Ulijaszek, S. J. (1992b). Dietary and nutrient intakes of 25 Ningerum (New Guinea) adult males at two times of the year. *American Journal of Human Biology*, 4, 469–79.

Ulijaszek, S. J. (1993a). Influence of birth interval and child labour on family energy requirements and dependency ratios in two traditional subsistence economies in Africa. *Journal of Biosocial Science*, 25, 79–86.

Ulijaszek, S. J. (1993b). Seasonality of reproductive performance in rural Gambia. In *Seasonality and Human Ecology*. ed. S. J. Ulijaszek & S. S. Strickland, pp. 76–88. Cambridge: Cambridge University Press.

Ulijaszek, S. J. (1994). Between-population variation in pre-adolescent growth. *European Journal of Clinical Nutrition*, 48, Supplement 1, S1–13.

Ulijaszek, S. J. (1995). Energetics, adaptation, and adaptability. *American Journal of Human Biology*, in press.

Ulijaszek, S. J. & Lourie, J. A. (1994). Intra- and inter-observer error in anthropometric measurement. In *Anthropometry: the Individual and the Population*. ed. S. J. Ulijaszek & C. G. N. Mascie-Taylor, pp. 30–55. Cambridge: Cambridge University Press.

Ulijaszek, S. J. & Mascie-Taylor, C. G. N. (eds.) (1994). *Anthropometry: the Individual and the Population*. Cambridge: Cambridge University Press.

Ulijaszek, S. J. & Poraituk, S. P. (1983). Subsistence patterns and sago cultivation in the Purari delta. In *The Purari – Tropical Environment of a High Rainfall River Basin*. ed. T. Petr, pp. 577–88. The Hague: Dr W. Junk Publishers.

Ulijaszek, S. J. & Poraituk, S. P. (1993). Making sago in Papua New Guinea: is it worth the effort? In *Tropical Forests, People and Food. Biocultural Interactions and Applications to Development*. ed. C. M. Hladik, A. Hladik, O. F. Linares, H. Pagezy, A. Semple & M. Hadley, pp. 271–80. Paris: UNESCO.

Ulijaszek, S. J. & Strickland, S. S. (1991). Basal metabolic rate and physique of Gurkha and British soldiers stationed in Britain. *Annals of Human Biology*, 18,

245–51.

Ulijaszek, S. J. & Strickland, S. S. (1993a). *Nutritional Anthropology. Prospects and Perspectives.* London: Smith-Gordon.

Ulijaszek, S. J. & Strickland, S. S. (1993b). Nutritional studies in biological anthropology. In: *Research Methods in Biological Anthropology.* ed. C. G. N. Mascie-Taylor & G. Lasker, pp. 108–39. Cambridge: Cambridge University Press.

Ulijaszek, S. J., Hyndman, D. C., Lourie, J. A. & Pumuye, A. (1987). Mining, modernisation and dietary change among the Wopkaimin of Papua New Guinea. *Ecology of Food and Nutrition,* 20, 148–56.

Ulijaszek, S. J., Brown, T. & Lourie, J. A. (1993). Physical work capacity, body size and within-household variation in energy and protein intakes in the North Fly Region of Papua New Guinea. *Man and Culture in Oceania,* 10, 1–10.

Ulijaszek, S. J., Tuffrey, V. R. & Strickland, S. S. (1994). Male–female differences in total energy expenditure and activity levels in Nepal. *American Journal of Physical Anthropology,* Supplement 18, 199–200.

Underwood, L., Smith, E. P., Clemmons, D. R., Maes, M., Maiter, D. & Ketelslegers, J-M. (1989). The production and actions of insulin-like growth factors: their relationship to nutrition and growth. In *Auxology 88. Perspectives in the Science of Growth and Development.* ed. J. M. Tanner, pp. 235–49. London: Smith-Gordon.

United Nations Children's Fund (1990). *The State of the World's Children.* Oxford: Oxford University Press.

United States Department of Health, Education, and Welfare (1972). *Food Composition Table for Use in East Asia.* Rome: Food and Agriculture Organisation of the United Nations.

Vasquez-Velasquez, L. (1988). Energy expenditure and physical activity of malnourished Gambian infants. *Proceedings of the Nutrition Society,* 47, 233–9.

Vasquez-Velasquez, L. (1989). Energy expenditure of infants. PhD thesis. Cambridge: Cambridge University Press.

Vaughan, L., Zurlo, F. & Ravussin, E. (1991). Aging and energy expenditure. *American Journal of Clinical Nutrition,* 53, 821–5.

Veldhuis, J. D., Evans, W. S., Demers, L. M., Thorner, M. O., Wakat, D. & Rogol, A. D. (1985). Altered neuroendocrine regulation of gonadotrophin secretion in women distance runners. *Journal of Clinical Endocrinology and Metabolism,* 6, 557–63.

Vickers, R. (1988). Effectiveness of defenses: a significant predictor of cortisol excretion under stress. *Journal of Psychosomatic Research,* 32, 21–9.

Villar, J. & Belizan, J. (1982a). The relative contribution of prematurity and fetal growth retardation to low birthweight in developing and developed societies. *American Journal of Obstetrics and Gynecology,* 143, 793–8.

Villar, J. & Belizan, J. (1982b). The timing factor in the pathophysiology of the intrauterine growth retardation syndrome. *Obstetric and Gynecological Survey,* 37, 499–506.

Villar, J., Belizan, J., Spalding, J. & Klein, R. (1982). Postnatal growth of intrauterine growth retarded infants. *Early Human Development,* 6, 265–71.

Villar, J., Smeriglio, V., Martorell, R., Brown, C. & Klein, R. (1984). Heterogeneous growth and mental development of intrauterine growth-retarded infants during the first 3 years of life. *Pediatrics*, **64**, 783–91.

Viteri, F. E., Torun, B., Garcia, J. C. & Herrera, E. (1971). Determining energy costs of agricultural activities by respirometer and energy balance techniques. *American Journal of Clinical Nutrition*, **24**, 1418–30.

Waddell, E. (1972). The mound builders: agricultural practices, environment and social in the central highlands of New Guinea. *American Ethnological Society Monograph No. 53*. Seattle: University of Washington Press.

Walsh, J. A. & Warren, K. S. (1979). Selective primary health care. An interim strategy for disease control in developing countries. *New England Journal of Medicine*, **301**, 967–74.

Ware, H. (1979). Social influences on fertility at later ages of reproduction. *Journal of Biosocial Science*, Supplement, **6**, 75–96.

Warren, M. P. (1990). Weight control *Seminars in Reproductive Endocrinology*, **8**, 25–31.

Waterlow, J. C. (1986). Notes on the new estimates of energy requirements. *Proceedings of the Nutrition Society*, **45**, 351–60.

Waterlow, J. C. (1988). Observations on the natural history of stunting. In *Linear Growth Retardation in Less Developed Countries*, ed. J. C. Waterlow, pp. 1–12. New York: Raven Press.

Waterlow, J. C. (1990a). Energy sparing mechanisms: reductions in body mass, BMR and activity: their relative importance and priority in undernourished infants and children. In *Activity, Energy Expenditure and Energy Requirements of Infants and Children*. ed. B. Schurch & N. S. Scrimshaw, pp. 239–50. Lausanne: International Dietary Energy Consultancy Group.

Waterlow, J. C. (1990b). Mechanisms of adaptation to low energy intakes. In *Diet and Disease*. ed. G. A. Harrison & J. C. Waterlow, pp. 5–23. Cambridge: Cambridge University Press.

Waterlow, J. C., Ashworth, A. & Griffiths, M. (1980). Faltering in infant growth in less developed countries. *Lancet* **ii**, 1176.

Watts, M. (1986). Drought, environment, and food security: some reflections on peasants, pastoralists, and commoditization in dryland West Africa. In *Drought and Hunger in Africa*. ed. M. H. Glantz, pp. 171–212. Cambridge: Cambridge University Press.

Watts, M. (1988). Coping with the market: uncertainty and food security among Hausa peasants. In *Coping with Uncertainty in Food Supply*. ed. I. de Garine and G. A. Harrison, pp. 260–89. Oxford: Clarendon Press.

Webb, P. (1986). 24-hour energy expenditure and the menstrual cycle. *American Journal of Clinical Nutrition*, **44**, 614–19.

Weiner, J. S. (1972). Tropical ecology and population structure. In *The Structure of Human Populations*. ed. G. A. Harrison & A. J. Boyce, pp. 72–86. Oxford: Clarendon Press.

Weiner, J. S. & Lourie, J. A. (1981). *Practical Human Biology*. London: Academic Press.

Weitz, C. A., Greksa, L. P., Thomas, R. B. & Beall, C. M. (1989). An anthropological perspective on the study of work capacity. In *Human Population Biology*. ed. M. A. Little & J. D. Haas, pp. 113–31. Oxford: Oxford University Press.

Westerterp, K. R., Saris, W. H. M., van Es, M. & ten Hoor, F. (1986). Use of the doubly labelled water technique in humans during heavy sustained exercise. *Journal of Applied Physiology*, **61**, 2162–7.

Westerterp, K. R., Brouns, F., Saris, W. H. M. & ten Hoor, F. (1988). Comparison of doubly labelled water with respirometry at low- and high-activity levels. *Journal of Applied Physiology*, **65**, 53–6.

Wheeler, E. F. & Abdullah, M. (1988). Food allocation within the family: response to fluctuating food supply and food needs. In *Coping with Uncertainty in Food Supply*. ed. I. de Garine & G. A. Harrison, pp. 437–51. Oxford: Oxford University Press.

Wheeler, P. E. (1991). The influence of bipedalism on the energy and water budgets of early hominids. *Journal of Human Evolution*, **20**, 117–36.

Wheeler, P. E. (1992). The influence of the loss of functional body hair on the energy and water budgets of early hominids. *Journal of Human Evolution*, **23**, 379–88.

Wheeler, P. E. (1993). The influence of stature and body form on hominid energy and water budgets; a comparison of *Australopithecus* and early *Homo* physiques. *Journal of Human Evolution*, **24**, 13–28.

White, L. A. (1949). *The Science of Culture: A Study of Man and Civilization*. New York: Grove Press.

White, L. A. (1959). *The Evolution of Culture*. New York: McGraw-Hill.

Whitehead, R. G. (1985). Infant physiology, nutritional requirements, and lactational adequacy. *American Journal of Clinical Nutrition*, **41**, 447–58.

Whitehead, R. G. & Paul, A. A. (1988). Comparative infant nutrition in man and other animals. In *Comparative Nutrition*. ed. K. Blaxter & I. MacDonald, pp. 199–213. London: John Libbey.

Whitehead, R. G., Rowland, M. G. M., Hulton, M., Prentice, A. M., Muller, E. & Paul, A. (1978). Factors influencing lactation performance in rural Gambian mothers. *Lancet* **ii**, 178–81.

Wilke, R. R. & Netting, R. Mc C. (1984). Households: changing forms and functions. In *Households: Comparative and Historical Studies of the Domestic Group*. ed. R. Netting, R. Wilke & E. J. Arnould, pp. 1–28. Berkeley: University of California Press.

Willett, W. C. (1990). *Nutritional Epidemiology*. Oxford: Oxford University Press.

Willett, W. C., Sampson, L., Stampfer, M. J. *et al*. (1985). Reproductibility and validity of a semiquantitative food frequency questionnaire. *American Journal of Epidemiology*, **122**, 51–65.

Williams, D. P., Going, S. B., Lohman, T. G., Hewitt, M. J. & Haber, A. E. (1992). Estimation of body fat from skinfold thicknesses in middle-aged and older men and women: a multiple component approach. *American Journal of Human Biology*, **4**, 595–605.

Williams, R. S. (1986). Regulation of mitochondrial biogenesis by contractile activity in skeletal muscle: recent advances and directions for future research. In *Biochemical Aspects of Physical Exercise*. ed. G. Benzi, L. Packer & N. Siliprandi, pp. 101–12. Amsterdam: Elsevier.

Wilmsen, E. (1978). Seasonal effects of dietary intake on Kalahari San. *Federation Proceedings*, **37**, 65–72.

Winterhalder, B. (1984). Reconsidering the ecosystem concept. *Reviews in Anthropology*, **11**, 301–30.

Wising, P. J. (1934). *Acta Medica Scandinavica*, **81**, 487.

Wood, J. W., Johnson, P. L. & Campbell, K. L. (1985). Demographic and endocrinological aspects of low natural fertility in highland New Guinea. *Journal of Biosocial Science*, **17**, 57–79.

Woodburn, J. (1972). Ecology, nomadic movement and the composition of the local group among hunters and gatherers: an East African example and its implications. In *Man, Settlement and Urbanism*. ed. P. J. Ucko, R. Tringham & G. W. Dimbleby. London: Duckworth.

Wyon, J. B., Finner, S. L. & Gordon, J. E. (1966). Differential age at menopause in rural Punjab, India. *Population Index*, **32**, 328.

Yambi, O. (1988). Nutritional status and the risk of death: a prospective study of children six to thirty months old in Ivinga Region, Tanzania. Ph.D. thesis, Cornell University.

Yarnell, J. W. G., Fehily, A. M., Milbank, J. E., Sweetnam, P. M. & Walker, C. L. (1983). A short questionnaire for use in epidemiological surveys: comparison with weighed dietary records. *Human Nutrition: Applied Nutrition*, **37A**, 103–12.

Zumrawi, F. Y., Dimond, H. & Waterlow, J. C. (1987). Effects of infection on growth in Sudanese children. *Human Nutrition: Clinical Nutrition*, **41C**, 453–61.

Index